INDUCTION, PROBABILITY, AND CAUSATION

SYNTHESE LIBRARY

MONOGRAPHS ON EPISTEMOLOGY,

LOGIC, METHODOLOGY, PHILOSOPHY OF SCIENCE,

SOCIOLOGY OF SCIENCE AND OF KNOWLEDGE,

AND ON THE MATHEMATICAL METHODS OF

SOCIAL AND BEHAVIORAL SCIENCES

Editors:

DONALD DAVIDSON, *Princeton University*

JAAKKO HINTIKKA, *University of Helsinki and Stanford University*

GABRIËL NUCHELMANS, *University of Leyden*

WESLEY C. SALMON, *Indiana University*

HUMANITIES PRESS / NEW YORK

INDUCTION, PROBABILITY,
AND CAUSATION

Selected Papers by C. D. Broad

D. REIDEL PUBLISHING COMPANY / DORDRECHT-HOLLAND

SOLE DISTRIBUTORS FOR U.S.A. AND CANADA
HUMANITIES PRESS / NEW YORK

Library of Congress Catalog Card Number 68-25475

1968

All rights reserved
No part of this book may be reproduced in any form, by print, photoprint, microfilm, or any other means, without permission from the publisher

Printed in The Netherlands by D. Reidel, Dordrecht

EDITORIAL PREFACE

In his essay on 'Broad on Induction and Probability' (first published in 1959, reprinted in this volume), Professor G. H. von Wright writes: "If Broad's writings on induction have remained less known than some of his other contributions to philosophy..., *one* reason for this is that Broad never has published a book on the subject. It is very much to be hoped that, for the benefit of future students, Broad's chief papers on induction and probability will be collected in a single volume...."

The present volume attempts to perform this service to future students of induction and probability. The suggestion of publishing a volume of this kind in Synthese Library was first made by Professor Donald Davidson, one of the editors of the Library, and was partly prompted by Professor von Wright's statement. In carrying out this suggestion, the editors of Synthese Library have had the generous support of Professor Broad who has among other things supplied a new Addendum to 'The Principles of Problematic Induction' and corrected a number of misprints found in the first printings of this paper. The editors gratefully acknowledge Professor Broad's help and encouragement.

A bibliography of Professor Broad's writings (up to 1959) has been compiled by Dr. C. Lewy and has appeared in P. A. Schilpp, editor, *The Philosophy of C. D. Broad* (The Library of Living Philosophers), pp. 833–852. A brief list of Broad's writings on induction and probability is found in Professor von Wright's article (pp. 228–229 *infra*). As can be seen from a comparison between this list and the table of contents of the present volume, all Broad's systematic papers on induction and probability are reprinted here.

In addition to them, Professor Broad has published in *Mind* a series of important critical notices or review articles in this area. In the words of Professor von Wright, these "may be said to constitute a running commentary on the developments in inductive logic from the appearance of

EDITORIAL PREFACE

Keynes' *Treatise* in 1921 to Kneale's *Probability and Induction* in 1949." These review articles contain many highly interesting comments on and proposals for further development of the ideas put forward in the work to be reviewed. Because of their interest, substantial parts of these review articles are reprinted in the present volume. The omitted parts consist mainly of Broad's summaries of parts of the books under review. Masterful as these summaries are, it did not seem motivated to reprint all of them here in view of the ready accessibility of these books themselves. However, Broad's Critical Notices of Keynes, von Mises and Kneale are reprinted here almost entirely unabridged. In fact, the omitted parts are very brief, mainly such as have been made out of date by subsequent developments. In the case of Broad's three-part review article of G.H. von Wright's *The Logical Problem of Induction*, the review dealt with the first edition of a work which has since been substantially revised, in some cases apparently because of Broad's comments. This seemed to motivate the exclusion of this review article. The only other essay by Professor Broad on induction and probability of which no part is reprinted here is the famous essay *The Philosophy of Francis Bacon*, which was published as a separate pamphlet by Cambridge University Press in 1936. It has been reprinted in a collection of Professor Broad's papers entitled *Ethics and the History of Philosophy* (Routledge and Kegan Paul, London, 1952), and is therefore easily available.

Neither Professor Broad nor his philosophical work needs a special introduction. An interested reader will find a good starting point in the volume on Broad in the Library of Living Philosophers which was mentioned above. Broad's views on induction and probability are discussed in some detail by G.H. von Wright in the paper 'Broad on Induction and Probability' which has already been referred to and which is reprinted in this volume. It is given additional interest by Broad's own comments on it which were published as a part of Broad's 'Reply to my Critics' in P.A. Schilpp, editor, *The Philosophy of C.D. Broad*, pp. 745–764. These comments are reproduced in the present volume, together with some comments by Professor Broad on the other papers in the Schilpp volume.

All the papers are reprinted here with the appropriate permission, in the case of the material from the Schilpp volume by courtesy of The Open Court Publishing Company, in the case of the papers 'The Principles of Problematic Induction' (*Proc. of the Aristotelian Society*, N.S. **28**

EDITORIAL PREFACE

(1927–28) 1–46), and 'Mechanical and Teleological Causation' (*ibid.*, Supplementary Volume **14** (1935) 83–112) by courtesy of the Editor of the Proceedings of The Aristotelian Society, and in the case of all the other material by courtesy of the Editor of *Mind*. It is understood that the International Copyrights to these different papers are not affected by their being reprinted here.

No substantive changes are made in any of the papers. Misprints have been corrected, some bibliographical references have been expanded, and a few editorial footnotes have been supplied. Professor Broad's symbolism and notation has been preserved throughout, although the style of some of the bibliographical references is slightly modified. Omissions have always been indicated either by the customary dots or by a footnote.

It is our hope that this volume fulfills the hopes expressed by Professor von Wright in our first quotation.

JAAKKO HINTIKKA
For the Editors

PREFACE

This volume, as Professor Hintikka explains in the Editorial Preface, contains by far the greater part of what I have published in the course of my life in various scattered places on the closely interconnected topics of Induction, Probability, and Causation. It excludes, for the good reasons stated by him, two and only two of such writings, viz., *The Philosophy of Francis Bacon* (Cambridge University Press, 1926) and the three articles in *Mind* 1944 entitled 'Hr. von Wright on the Logic of Induction'.

I am very greatly indebted to Professor Hintikka, and to any of his colleagues who may have been concerned, for the inception of the idea and for the time and care which they have devoted to carrying it to completion.

In expressing my feelings of gratitude I would add a special word of thanks to Professor von Wright, a friend of long standing. His contribution 'Broad on Induction and Probability', reprinted from *The Philosophy of C. D. Broad* in Schilpp's *Library of Living Philosophers*, is an outstanding feature in the present book. One could not hope to have a fuller or a fairer critical synopsis of one's writings than this which von Wright, who is an expert on their subject-matter, here provides. Any reader of the present volume would do well to begin by reading von Wright's contribution, in order to get a good critical idea of the contents of the rest of the book. He can then turn to those parts of the volume (if any) which may seem to him worth his further detailed consideration.

In this connexion I would add that I am very pleased that the relevant parts of my 'Reply to my Critics' in the Schilpp volume are here reproduced. That volume appeared in 1959, and the 'Reply' in question was completed towards the end of 1956. I hope that it may throw light on some matters which were obscure to von Wright, and also that it may help to clear up certain other obscurities in my writings.

I notice that the earliest of my articles was originally printed in 1918, when I was in my 31st year. The latest, viz., the 'Reply' from the Schilpp volume, was, as I have said above, written in 1956, when I was in my 69th year. Most of the others date from my early 30's to my late 50's. As I write this *Preface* I am in the first quarter of my 81st year.

After this it needs scarcely to be emphasized that the style and the content of these writings will strike present-day readers, if any, as extremely old-fashioned. There is not a word in them of 'talk about "talk"'! Fashions in philosophy change very quickly, and there have been quite a number of them during the period of some 40 years covered by the contents of the book. I can only hope that some of the essays go far enough back to have begun, like Victorian furniture, to acquire something of the patina of 'period pieces'.

C. D. BROAD

CONTENTS

The Relation between Induction and Probability I–II	1
A Critical Notice of J. M. Keynes, *A Treatise on Probability*	53
A Critical Notice of W. E. Johnson, *Logic*, Part II (abridged)	69
Mr. Johnson on the Logical Foundations of Science I–II (abridged)	73
The Principles of Problematic Induction (with an Addendum)	86
The Principles of Demonstrative Induction	127
Mechanical and Teleological Causation	159
A Critical Notice of R. von Mises, *Wahrscheinlichkeit, Statistik, und Wahrheit*	184
A Critical Notice of Wm. Kneale, *Probability and Induction*	201
GEORG HENRIK VON WRIGHT / Broad on Induction and Probability	228
C. D. BROAD / Replies to my Critics	273

THE RELATION BETWEEN INDUCTION
AND PROBABILITY

PART I

In the present paper I propose to try to prove three points, which, if they can be established, are of great importance to the logic of inductive inference. They are (1) that *unless* inductive conclusions be expressed in terms of probability all inductive inference involves a formal fallacy; (2) that the degree of belief which we actually attach to the conclusions of well-established inductions cannot be justified by any known principle of probability, unless some further premise about the physical world be assumed; and (3) that it is extremely difficult to state this premise so that it shall be at once plausible and non-tautologous. I believe that the first two points can be rigorously established without entering in detail into the difficult problem of what it is that probability-fractions actually measure. The third point is more doubtful, and I do not profess to have reached at present any satisfactory view about it.

1

All inductions, however advanced and complicated they may be, ultimately rest on induction by simple enumeration or on the use of the hypothetical method. We shall see at a later stage the precise connexion between these two methods. In the meanwhile it is sufficient for us to notice that, whilst the inductions of all advanced sciences make great use of deduction, they can never be reduced without residue to that process. In working out the consequences of a scientific hypothesis many natural laws are assumed as already established and much purely deductive reasoning is used. But the evidence for the assumed laws will itself be ultimately inductive, and the use which is made of our deduced conclusions to establish the hypotheses by their agreement with observable facts involves an inductive argument.

Now both induction by simple enumeration and the hypothetical method involve, on the face of them, formal fallacies. The type of argument in the first kind of induction is: All observed S's have been P, therefore all S's whatever will be P. Now the observed S's are not known to be all the S's (indeed they are generally believed not to be all the S's). Hence we are arguing from a premise about *some* S's to a conclusion about *all* S's, and are clearly committing an illicit process of S.

Most inductive logicians of course recognise this fact, but most of them seem to suppose that the fallacy can be avoided by the introduction of an additional premise which they call the Uniformity of Nature or the Law of Causation. They admit that there is a difficulty in stating this principle satisfactorily and in deciding on the nature of the evidence for it, but they seem to feel no doubt that *if* it could be satisfactorily stated and established the apparent formal fallacy in induction by simple enumeration would vanish. It is easy, however, to show that this is a complete mistake. Whatever the supposed principle may be, and however it may be established, it cannot be stronger than an universal proposition. But if an universal proposition be added to our premise, All observed S's are P, the latter premise still remains particular as regards S. And from a universal and a particular premise no universal conclusion can be drawn.

It follows then that no additional premise, whether about logic or about nature, can save induction by simple enumeration from a formal fallacy, so long as the conclusion is in the form all S's are P. If the validity of the process is to be saved at all it can only be saved by modifying the conclusion. It remains of course perfectly possible that some additional premise about nature is *necessary* to justify induction; but it is certain that no such premise is *sufficient* to justify it.

The hypothetical method equally involves, on the face of it, a formal fallacy. The general form of the argument here is: If h be true then $c_1, c_2 \ldots c_n$ must be true. But $c_1, c_2 \ldots c_n$ are all found by observation to be true, hence h is true. This argument of course commits the formal fallacy of asserting the consequent in a hypothetical syllogism. The only additional premise which could validate such an argument would be the proposition: h is the only possible hypothesis which implies $c_1, c_2 \ldots c_n$. But this proposition is never known to be true and is generally known to be false.

The conclusions of inductive argument must therefore be modified, and the most reasonable modification to make is to state them in terms of probability. The advantages of such a course are (a) that this accords with what we actually believe when we reflect. We always admit that the opposite of an inductive conclusion remains possible; even when we *say* that such conclusions are certain we only mean that they are so probable that for all practical purposes we can act as if they were certain. That this differs from genuine certainty may be seen if we reflect on the difference in our attitude towards the true propositions, All grass is green, and $2 \times 2 = 4$. In ordinary language both would be called 'certain', but our attitudes towards the two are quite different. No one would care to assert that there might not be in some part of space or time something answering to our definition of grass but having a blue instead of a green colour.

(b) With the suggested modification of our conclusion the logical difficulty vanishes. Suppose the conclusion becomes: It is highly probable on the observed data that all S's are P. There is then no illicit process. We argue from a *certain proposition* about *some* S's to the *probability* of a proposition about *all* S's. This is perfectly legitimate. The subject of our conclusion is no longer All S's, but is the proposition All S's are P. The predicate is no longer P, but is the complex predicate 'highly probable with respect to the observed data'.

(c) If inductions with their unmodified conclusions were valid forms of inference we should be faced by a strange paradox which furnishes an additional proof that inductive conclusions must be modified. It is often certain that all the observed S's are P. Now what follows from a certain premise by a valid process of reasoning can be asserted by itself as true. Yet we know quite well that, if the conclusion of an inductive argument be All S's are P, the very next observation that we make may prove this conclusion to be false. Hence we have the paradox that, if induction be valid and the conclusion be All S is P, a *certain* premise and a *valid* argument may lead to a *false* conclusion. This paradox is removed if we modify our conclusion to the form: It is highly probable on the observed data that all S is P. Probability and truth-value are both attributes of propositions. (I omit here further subtleties as to whether they do not more properly belong to propositional forms, or, as Russell calls them, functions.) But they are very different attributes. (i) A proposition is true or false in itself and without regard to its relations to other propositions; a proposition has probability only with respect to others, and it has different probabilities with respect to different sets of data. (ii) A proposition which is very probable with respect to certain data may be in fact false, and conversely. This is precisely what we mean by 'a strange coincidence'. It follows from these facts that if I have observed n S's and they were all P it may be highly probable relative to these data that all S's are P, and yet it may be false that all S is P. If I observe an $n+1$th S and it proves not to be P, I know that it is false that all S is P; but this does not alter the truth of the proposition that, relative to my first n observations, it *is* highly probable that all S is P. For the probability of a proposition may be high with respect to one set of data and may be zero with respect to another set which includes the former. Our original inductive conclusion does not cease to be *true*, it only ceases to be practically important.

For all these reasons I hold that we have established the point that inductive *conclusions* must be modified if induction is to be saved and that no additional *premises* will suffice to save it. And I think it almost certain that the direction in which the modification must be made is the one which I have indicated above. Leibniz said in a famous passage that Spinoza would be right if it were not for the monads; we may say that Hume would be right if it were not for the laws of probability. And just as it is doubtful whether Leibniz was right even with the monads, so there remains a grave doubt whether induction can be logically justified even with the laws of probability.

2

If we accept the view that inductive conclusions are in terms of probability, it is clear that a necessary premise or principle of all inductive argument will be some proposition or propositions concerning probability. Since probability, like truth, implication, etc., is an attribute of propositions, the laws of probability are laws of logic, not of nature, just like the principle of the syllogism or the law of contradiction. That is, they are principles which hold in all possible worlds, and do not depend on the special structure of the world that actually exists. It remains possible however that they are only capable of fruitful application to real problems if the actual world fulfils certain conditions which need not be fulfilled in all possible worlds. *E.g.* $2 \times 2 = 4$ holds in all possible worlds, but it would be very difficult to make any practical use of this proposition in

THE RELATION BETWEEN INDUCTION AND PROBABILITY

physics if all objects in the actual world were like drops of water and ran together into a single drop when mixed.

To see what the principles of probability required by induction are, and to consider whether they suffice to justify the actual strength of our beliefs in universal propositions about matters of fact, I propose to consider induction by simple enumeration and the hypothetical method in turn.

A. *Induction by simple enumeration*

The way in which I propose to treat this problem is as follows. I shall first consider the logical principles employed and the factual assumptions made when we draw counters out of a bag, and, finding that all which we have drawn are white, argue to the probability of the proposition that all in the bag are white. I shall then discuss as carefully as I can the analogies and differences between this artificial case and the attempt to establish laws of nature by induction by simple enumeration. We shall then be able to see whether an alleged law of nature can logically acquire a high degree of probability by this method, and, if not, what additional assumptions are needed.

We will divide the factors of the problem into three parts, (a) Facts given, (b) Principles of probability accepted as self-evident, (c) Factual assumptions made.

(a) The facts given are:

(i) That the bag contains n counters indistinguishable to touch.

(ii) That we have no information at the outset of the experiment what proportion of the counters are white; there may be $0, 1, 2, \ldots n$ whites. (We know of course on *a priori* grounds that any one proportion, so long as it subsists, excludes any other, and that, at any given moment, one of these $n+1$ possible proportions must subsist.)

(iii) That at the end of the experiment m counters have been drawn out in succession, none being replaced, and that these have all been found to be white.

(b) The principles of probability accepted as *a priori* truths are:

(i) If p and q be two mutually exclusive propositions and $x|h$ means 'the probability of x given h', then

$$p \vee q|h = p|h + q|h.$$

(ii) If p and q be any two propositions, then

$$p.q|h = p|h \times q|p.h = q|h \times p|q.h.$$

(iii) If we know that several mutually exclusive alternatives are possible and do not know of any reason why one rather than another should be actual, the probability of any one alternative, relative to this precise state of knowledge and ignorance, is equal to that of any other of them, relative to the same data.

(iv) The present proposition is to be regarded rather as a convention for measuring probability than as a substantial proposition. It is: If p and q be coexhaustive and coexclusive propositions, then

$$p|h + q|h = 1.$$

(c) The assumptions which we make about matters of fact are:

(i) That in drawing out a counter our hand is as likely to come in contact with any one as with any other of all those present in the bag at the moment.

(ii) That no process going on in nature during the experiment alters the total number or the proportion of the white counters, and that the constitution of the contents only changes during the experiment by the successive removal of counters by the experimenter.

It is clear that the propositions (c) are assumptions about the course of nature and have no *a priori* guarantee. This is perfectly obvious about c (ii), and it is evident that a factual assumption is an essential part of c (i) even if the *a priori* factor b (iii) should also somewhere be involved in it.

On these assumptions it can be proved that the probability that the *next* to be drawn will be white is $\frac{m+1}{m+2}$, and that the probability that *all* the n are white is $\frac{m+1}{n+1}$. I do not propose to go into the details of the argument, which involves the summation of two series. What I wish to point out is that all the nine propositions mentioned above are used in the proof and that no others are needed except the ordinary laws of logic and algebra. It is easy to see in a general way how the assumptions (c) enter. Suppose there were a kind of pocket in the bag and that non-whites happened to be accumulated there. Then c (i) would be false, and it is clear that a large number of whites might be drawn at first and give a misleadingly high expectation of all being white even though there were quite a large proportion of non-whites in the bag. Suppose again that c (ii) were false and that the proportion of whites might change between one draw and the next. Putting the course of the argument very roughly indeed we may say that at the beginning we start with $n+1$ equally likely hypotheses as to the constitution of the bag's contents. As we go on drawing whites and no non-whites we learn more of this constitution, certain of these hypotheses are ruled out altogether, the others have their probabilities strengthened in various degrees. But this is only true if we really do learn more about the constitution of the contents by our successive drawings; if, between these, the constitution changes from any cause, we have learnt nothing and the argument breaks down.

We can now consider how far the attempt to establish laws of nature by simple enumeration is parallel to the artificial example just dealt with. For clearness it is best to distinguish here between laws about the qualities of classes of substances [such as the law that All crows are black] and laws about the connexion of events [such as All rises of temperature are followed by expansion]. I do not suggest that this distinction is of great philosophic importance or is ultimately tenable, but it will help us for the present.

There is obviously a very close analogy between investigating the colours of crows and the colours of the counters in a bag. To the counters in the bag correspond all the crows in the universe, past, present, and future. To the pulling out and observing the colour of a counter corresponds the noticing of a certain number of crows. At this point however, the analogy fails in several ways, and all these failures tend to reduce the probability of the suggested law. (i) The same crow might be observed several times over and counted as a different one. Thus m in the fraction $\frac{m+1}{n+1}$ might be counted to be larger than it really is and the probability thus overestimated. (ii) We have no guarantee whatever that crows may not change their colours in the course of their lives. (This possibility was of course also present in the artificial case of counters, and our only ground for rejecting it is in previous inductions.) (iii) It is quite certain that we are not equally likely to meet with any crow. Even if we grant that any past crow is equally likely to have been met with and its colour reported to us, we know that the assumption of equiprobability is false as to future crows.

For we clearly cannot have observed any of the crows that begin to exist after the moment when we make the last observation which we take into account when we make our induction. And the assumption of equiprobability is most precarious even as regards past and present crows. Neither by direct observation nor by the reports of others can I know about crows in any but a restricted region of space. Thus the blackness of the observed crows may not be an attribute of all crows but may be true only of crows in a certain area. Outside this it may fail, as whiteness has been found to fail in the case of Australian swans. Our situation then is like that which would arise with the bag of counters if (a) there were a rigid partition in it past which we could not get our hands (distinction of past and future cases), and (b) if the bag were much bigger than the extreme stretch of our arm and we could only enter it through one comparatively small opening (restricted area of observation in space). We may sum up this objection by saying that the argument which leads to the probability $\frac{m+1}{n+1}$ assumes that a 'fair selection' has been observed, and that in the case of the crows we know that a 'fair selection' cannot have been observed owing to the fact that I cannot *now* observe *future* instances, and that I cannot directly observe even contemporary instances in all parts of space.

It is easy to prove that when we know that a 'fair selection' has not been observed the probability of a general law must fall below and can never rise above the value $\frac{m+1}{n+1}$ which it reaches if the observed selection be a fair one. Let us suppose that all the S's that might actually have been observed were SQ's; that, *within this class*, the selection observed was a fair one, though not fair for the S's as a whole; and that the number of SQ's is v. Then, since the number of SQ's examined was m and all were found to be P, the probability that all SQ's are P is $\frac{m+1}{v+1}$. The number of S$\bar{Q}$'s is $n-v$; but, by hypothesis, none of these came under examination. Hence we have no information whatever about them, and the probability that any proportion from O to the whole $n-v$ inclusive is P is the same, viz., $\frac{1}{n-v+1}$. Now the probability that All S's are P = the probability of the compound proposition: All SQ's are P and All S$\bar{Q}$'s are P. This cannot

THE RELATION BETWEEN INDUCTION AND PROBABILITY

exceed $\dfrac{m+1}{v+1}\dfrac{1}{n-v+1}$. It is evident that this is less than $\dfrac{m+1}{n+1}$; for its numerator is the same, whilst its denominator is $n+1+v(n-v)$, which is greater than $n+1$, since v is a positive integer less than n.

(iv) Lastly there is the following fatal difference even if all other difficulties could be overcome. In investigating the counters in the bag we know the total number n. It is finite, and we can make the number m of counters observed approximate fairly closely to it. We do not of course know the total number of crows that have been, are, and will be; but we can be perfectly sure that it must be enormous compared with the number investigated. Hence m is very small compared with n in the investigation of any natural law. Hence $\dfrac{m+1}{n+1}$, the probability of the law, as determined by induction by simple enumeration, is vanishingly small even under the impossibly favourable conditions of a 'fair selection'. In real life it will be indefinitely smaller than this indefinitely small fraction.

It must be noted, however, that from the same premises from which we deduced the expression $\dfrac{m+1}{n+1}$ for the probability that *all* S's are P we also deduced the expression $\dfrac{m+1}{m+2}$ for the probability that the *next* S to be examined will be P. A more general formula which can also be proved from the same premises is that the probability that the next μ to be examined will be P is $\dfrac{m+1}{m+\mu+1}$. These latter expressions, it will be noted, are independent of n. Hence, if we could get over the difficulties about a 'fair selection' and about possible changes in time and possible repeated examinations of the same S, induction by simple enumeration would play a modest but useful *rôle* even in the investigation of nature. If m were pretty large both in itself and as compared with μ we could predict for the next case and for the next few with tolerable certainty. But this assumes that the 'next case' is one which had as much likelihood as any other of falling under our observations, though it did not actually do so. In the case of persistent entities like counters and crows this condition may perfectly well be fulfilled, for the 'next' simply means the 'next which I happen to observe'. In the case of the counters the one which I shall pull out next was in the bag all through the experiment and was as likely to be taken out as those which actually were taken out. In that of the crows the crow that I shall next observe may have existed when I observed the previous ones, and may have been as likely to fall under my observation as any of those which actually did so. But, as we shall see in a moment, there are special difficulties about events which will not allow us to apply this reasoning to them.

We will now consider the connection of events. Much of what has been said about the investigation of the properties of substances remains true here, but there are the following differences to be noted. Suppose our events are rises in temperature. The class about which we wish to learn is all events of this kind past, present, and future. Now events, unlike substances, cannot change; each is tied to its own position in time and is determined by it. There is no possibility that the *same* rise in temperature should be at one moment followed by an expansion and at another not, as there is a possibility that the same crow may sometimes be black and sometimes white. Rises of temperature at different times are different rises of temperature; it is of course perfectly possible that one may be and another may not be followed by an expansion, but the *same* one cannot occur at two different moments and therefore cannot have different sequents at different times. Hence one difficulty inherent in investigating substances and their properties is ruled out in investigating events and their connexion.

For similar reasons there is no possibility of observing the same event twice, as there is of investigating the same crow twice. In observing events the position is quite parallel to pulling out counters and not putting them back. What is secured artificially in the counter experiment is secured in investigating events by the fact that each event is tied to its moment and ceases to belong to the class of observable events when that moment is past.

So far the inductive observation of events is in a stronger position than that of substances. But here its advantages cease. There is clearly the same impossibility of observing any finite proportion of the whole class, and hence of ascribing any appreciable probability to a general law about its members. There is the same difficulty about observing a 'fair selection' in space. And there is a still more hopeless difficulty about predicting the future even for the next event of the class. For it is perfectly certain that I could not up to now have observed any event which belongs to a moment later than my last observation. Hence the condition of equiprobability breaks down and my observations add nothing to the probability that the next event to be observed will agree with those which I have already observed. With substances, as we saw, it was possible that the next one to be observed had an equal chance of having been observed with any of those which I actually happened to notice. Hence there was a possibility of predicting a few steps ahead if we assume that the substances are not changing their qualities. But this is because substances persist for a time and are not tied to single moments like events.

I conclude then that, neither for substances nor for events, will the principles of probability alone allow us to ascribe a finite probability to general laws by induction by simple enumeration. In the case of substances we can argue a few steps ahead if we can assume a 'fair selection' in space, and can further assume that the substances do not change in the property in question over the time for which we are observing and predicting. For events even this amount of prediction is incapable of logical justification. And the latter fact really invalidates the process for substances. For, if our ground for assuming that the substances will not change their attributes be inductive, it must be an induction about events. The possession of an attribute at each moment of a duration constitutes a class of events, and to argue inductively that there will be no change is to argue from observations on some of the earlier members of this class to the later ones which cannot fall into the class of those which it was equally likely for us to have observed up to the moment at which we stop our observations. It was for this, among other reasons, that I said that the distinction between inductions about substances and inductions about events, though convenient in discussing the subject, was not of ultimate philosophic importance.

Before leaving induction by simple enumeration and passing to the hypothetical method it may be of interest to remark that, in theory, there are two quite different reasons for trying to enlarge the number of our observations as much as possible. (i) We want to examine as many S's as possible simply in order to increase the proportion of m to n in the fraction $\frac{m+1}{n+1}$. For this purpose it is quite irrelevant whether the observed instances happen under very similar or under very diverse circumstances. It is simply the number that counts. Unfortunately in investigating nature it is of little use to worry ourselves to increase m for this reason, since we know that however much we increase it, it will remain vanishingly small compared with n. (ii) We want to examine S's under as many different circumstances as possible. This is so as to approximate as nearly as we can to a 'fair selection'. Here it is not the mere number of instances that counts but the number of different circumstances under which the observed instances happen. Unfortunately however well we succeed in this we cannot raise the probability above $\frac{m+1}{n+1}$, we can only ensure that it shall not fall indefinitely below that indefinitely small fraction.

B. *The hypothetical method*

I shall first briefly state the connexion between this and induction by simple enumeration. I shall then consider the logical principles on which the hypothetical method is based and see whether they, without additional assumptions about nature, will suffice to give a finite probability to any suggested law.

Induction by simple enumeration is just a rather special case of the hypothetical method. At the outset of our experiment with the bag we have $n+1$ equally likely hypotheses as to the constitution of its contents. After the first draw has been made and the counter found to be white one of these hypotheses is definitely refuted (*viz.* that there were no whites present). The others remain possible but no longer equally probable; the probability of each on the new datum can be calculated. After the second draw another one hypothesis is definitely refuted; the remaining $n-1$ are all possible, but once more their probabilities have been altered in various calculable amounts by the addition of the new datum. The procedure after each draw (assuming that all turn out to be white) is the same; one hypothesis is always refuted; the rest always remain possible, and among these is always the hypothesis that all in the bag are white; and the probabilities of each are increased in various calculable degrees. The special peculiarities of this method are (a) that the various hypotheses are known to be mutually exclusive and to exhaust all the possibilities, (b) that they deal solely with the question of numbers or ratios, and (c) that only two of them, *viz.* the hypothesis that none are white and the hypothesis that all are white are comparable with general laws.

THE RELATION BETWEEN INDUCTION AND PROBABILITY

The reasoning of the hypothetical method in its most general form is the following. Let h be the hypothesis; it will consist of one or more propositions. We prove by ordinary deductive reasoning that h implies the propositions $c_1, c_2 \ldots c_n$. Let $h|f$ be the probability of the hypothesis relative to any data that we may have before we start our experiments to verify it. Then we know in general that

$$h.c_1|f = c_1|f \times h|c_1.f = h|f \times c_1|h.f.$$

If h implies c_1 it is clear that $c_1|h$ (and $\therefore c_1|h.f) = 1$.
Hence
$$c_1|f \times h|c_1.f = h|f.$$
Whence
$$h|c_1.f = \frac{h|f}{c_1|f}.$$
Again
$$h.c_1.c_2|f = h.c_1c_2|f = c_1c_2|f \times h|c_1c_2 f = h|f \times c_1c_2|h.f.$$
But
$$c_1c_2|hf = c_1|hf \times c_2|c_1hf = c_2|c_1hf.$$

And since h implies c_2 it is clear that $c_2|h$ (and $\therefore c_2|c_1hf) = 1$.
Hence
$$h|c_1c_2 f = \frac{h|f}{c_1c_2|f}.$$

In general, if h implies $c_1, c_2 \ldots c_n$, we shall have

$$h|c_1c_2 \ldots c_n f = \frac{h|f}{c_1c_2 \ldots c_n|f}.$$

We can learn much from a careful study of this formula. We see that the probability of a hypothesis is increased as we verify its consequences because the initial probability is the numerator of a fraction whose denominator is a product which contains more factors (and $\therefore$, since they are proper fractions, grows *smaller*) the more consequences we deduce and verify.

For $c_1c_2 \ldots c_n|f = c_1|f \times c_2|c_1 f \times c_3|c_2c_1 f \times \ldots \times c_n|c_{n-1} \ldots c_1 f$. Next we see that it is only by increasing the number of verified consequences which are logically independent of each other that we increase the probability

13

of the hypothesis. For if, e.g., c_{r-1} implies c_r the factor $c_r|c_{r-1}\ldots c_1 f = 1$ and so does nothing to reduce the denominator and thus increase the probability of the hypothesis. Again, the more unlikely the consequences were on the original data f which we had before we started to verify the hypothesis the more they increase the probability of the hypothesis if they be found to be true. For this means that the factors like $c_1|f$ are very small, hence that the denominator is small, hence that the final value of $h|c_1 c_2 \ldots c_n$ is likely to be large. This is the precise amount of truth that there is in the common view that an hypothesis is greatly strengthened by leading to some surprising consequence which is found to be true. The important point is not the psychological surprisingness of the consequence, but is the purely logical fact that *apart from* the hypothesis it was very unlikely to be true, *i.e.* it was neither implied nor rendered probable by anything that we knew when we put the hypothesis forward. Lastly we must notice that the factor $h|f$, expressing the probability of our hypothesis on the data known before any attempt at verification has been made, is always present in the numerator, *i.e.* as a multiplicative factor. Hence, unless we can be sure that this is not indefinitely small, we cannot be sure that the final probability of the hypothesis will be appreciable.

There is just one thing further to be said about $h|f$. h may be a complex set of propositions. Suppose we have two alternative hypotheses h_1 and h_2. Suppose $h_1 \equiv p_1 p_2 \ldots p_m$ and $h_2 \equiv q_1 q_2 \ldots q_n$, and let $n > m$. Then $h_2|f$ is a product of n factors all fractional and $h_1|f$ is a product of m factors all fractional. There will thus be a tendency for the less complex hypothesis to be more probable intrinsically than the more complex one. But this is only a tendency, not a general rule. The product $\frac{1}{2} \cdot \frac{3}{4} \cdot \frac{5}{6} \cdot \frac{7}{8}$ is greater than $\frac{1}{2} \cdot \frac{1}{3} \cdot \frac{5}{16}$, although the latter contains fewer factors than the former. This tendency, however, is the small amount of logical truth in the common notion that a more complicated hypothesis is less likely to be true than a simpler one.

We are now in a position to see whether the hypothetical method in general is any more capable of giving a finite probability to alleged laws of nature, without some additional premise, than its special case the method of induction by simple enumeration. I shall try to prove that, whilst the hypothetical method has many advantages which fully explain why it is the favourite instrument of all advanced sciences, it yet is

THE RELATION BETWEEN INDUCTION AND PROBABILITY

insufficient, without some further assumption, to establish reasonably probable laws.

The advantages of the method are obvious enough. (i) The hypotheses of induction by simple enumeration are purely numerical and therefore no consequence can be deduced from them except the probability of getting a certain number of favourable cases in a certain number of experiments. When hypotheses are not limited in this way the most varied consequences can be deduced, and, if verified, they increase the probability of the hypothesis. (ii) If the hypothesis be stated in mathematical form remote and obscure consequences can be deduced with all the certainty of mathematical reasoning. We thus have guidance as to what experiments to try, and powerful confirmation if our experiments succeed. The history of the wave theory of light is full of examples of this fact. (iii) If careful experiments refute some of the consequences of an hypothesis we know of course from formal logic that the hypothesis cannot, as it stands, be true. But if most of the deduced consequences have been verified we may fairly suspect that there cannot be much wrong with the hypothesis. And the very deductions which have failed to be verified may suggest to us the kind and degree of modification that is necessary. (iv) It is true that in induction by simple enumeration we have the advantage of knowing that our alternative hypotheses are exhaustive and exclusive. But in investigating nature this is of little profit since we also know that their number is indefinitely large. Now, it might be said, in the hypothetical method, although we cannot be sure that we have envisaged all possible alternatives, yet the number of possible laws to explain a given type of phenomena cannot be extremely great, hence the intrinsic probability of none of them will be excessively small if we regard them as all equally probable before attempted verification.

This last argument seems plausible enough at first sight. Yet it is mistaken, and in exposing the mistake we shall see why it is that the hypothetical method by itself will not give an appreciable probability to any suggested law. Why is it that the intrinsic probability of the law that all S is P is vanishingly small in induction by simple enumeration whilst that of any suggested law in the hypothetical method is not, to all appearance, vanishingly small? One reason is that the alternatives taken as intrinsically equally probable are not *in pari materia* in the two methods. In induction by simple enumeration the alternatives are not various possible *laws*, but various possible *proportions*, only two of which, *viz.* 0% and 100% of the S's being P, are laws. In the hypothetical method we have so far assumed that the alternative hypotheses are always laws. This naturally reduces the number of possible alternatives and hence increases the intrinsic probability of each as compared with the alternatives of induction by simple enumeration. But this difference renders comparison between the two methods unfair. If in simple enumeration alternatives other than laws are to be accepted as intrinsically as probable as laws there is no reason why the same assumption should not be made in the hypothetical method. And it is surely evident that the objections which apply to induction by simple enumeration as a sufficient means of establishing a law apply equally to the hypothetical method. All the experiments which have been made up to a given moment to verify an hypothesis can throw no light on the truth of this hypothesis as referring to moments after that at which the last experiment was performed. Now it is certain that an indefinite number of hypotheses could be put forward agreeing in their consequences up to a given moment and diverging after it. Exactly similar remarks apply to space; there can clearly be any number of alternative hypotheses which have the same consequences within a given region of space and different consequences outside it, and no experiments performed wholly within this region can give any ground for deciding between them. I think therefore that we may now claim to have proved our second contention that the degree of belief which we actually attach to the conclusions of well-established inductive arguments cannot be justified by any known principle of probability unless some further premise about the existent world be assumed. What this premise is, whether it can be stated clearly enough to admit of logical criticism, and whether in that event it will survive logical criticism, are extremely difficult questions which I reserve for the second part of this paper. What I have said so far I believe to be fairly certain, what I have yet to say I know to be extremely doubtful.

THE RELATION BETWEEN INDUCTION AND PROBABILITY

PART II

1

In the first part of this paper, I tried to show that the statement of inductive arguments in terms of probability is a necessary but not a sufficient condition of their validity. We saw that the laws of probability and the ordinary assumptions about equiprobability will not suffice to justify a strong belief in any law or even in a prediction for a few steps ahead. Some additional proposition about nature and not merely about probability seemed to be needed if induction were to be anything more than a guessing game in which we have so far had surprising luck. In this second part I propose to try and find what propositions are needed and what kind of evidence there is for them.

2

The usual view of the logic books seems to be that inductive arguments are really syllogisms with propositions summing up the relevant observations as minors, and a common major consisting of some universal proposition about nature. If this were true it ought to be easy enough to find the missing major, and the singular obscurity in which it is enshrouded would be quite inexplicable. It is reverently referred to by inductive logicians as the Uniformity of Nature; but, as it is either never stated at all or stated in such terms that it could not possibly do what is required of it, it appears to be the inductive equivalent of Mrs. Gamp's mysterious friend, and might be more appropriately termed Major Harris.

It is in fact easy to prove that this whole way of looking at inductive arguments is mistaken. On this view they are all syllogisms with a common major. Now their minors are propositions summing up the relevant observations. If the observations have been carefully made the minors are practically certain. Hence, if this theory were true, the conclusions of all inductive arguments in which the observations were equally carefully made would be equally probable. For what could vary their probabilities? Not the major, which is common to all of them. Not the minors, which, by hypothesis, are equally certain. Not the mode of reasoning, which is syllogistic in each case. But the result is preposterous, and is enough to refute the theory which leads to it.

Though we have thus cleared the ground of a false view its falsity leaves us with a much harder task than we should have had if it were true. For it is now by no means obvious in what direction to look for the missing premise about nature. Two courses seem open to us. (i) We might consider just where induction breaks down if it does not assume any premise about nature. We might then try to think of one or more propositions which would suffice to remove the difficulty. Lastly we might try to pare these down to their irreducible minimum and see whether they be self-evident or have any good evidence for or against them. (ii) But it will evidently be wise to use another method as a clue. We regard some inductive conclusions as fairly trustworthy and others as much less so. It will be wise to consider what assumptions or knowledge we have at the back of our minds when we make inductions. These may be betrayed by comparing the cases where we are satisfied with the induction with those where we are not. We can then state these assumptions explicitly; see whether they do suffice to make some inductions fairly probable; and consider the evidence for or against these assumptions. It seems reasonable to hope that the first method will suggest to us the *kind* of propositions about nature that are wanted, and that the second will suggest the actual propositions which people use when they make inductions. And we may hope that the latter will be instances of the former.

3

Induction by simple enumeration has so far been wrecked on two different reefs. (1) The number of S's examined could only bear a vanishingly small proportion to all the S's in the world, even if any one S were as likely to have fallen under our notice as any other. The result was that the number of antecedently equiprobable hypotheses about the proportion of S's which are P is enormous, and therefore the antecedent probability of the only pair which would be laws, *viz.*, All S is P and No S is P – is vanishingly small. (2) It is certain that not every S is equally likely to have fallen into the class of observed S's; for those which begin to exist after the experiment is concluded or exist in places remote from all the experimenters could not possibly have fallen into this class. It is pretty clear what *kind* of proposition is needed to diminish the first difficulty. We want some proposition which favours *laws* (*i.e.*, universal propositions) as against

THE RELATION BETWEEN INDUCTION AND PROBABILITY

propositions of the form $n\%$ of the S's in nature are P's; so that all S is P or no S is P shall be antecedently much more probable than the innumerable possible alternatives. And I have no doubt that this is what people must have had in mind when they spoke of the Uniformity of Nature and told us that it was a necessary premise of all inductions. But they hardly noticed how extremely difficult it is to state any such proposition in a form in which it is not flagrantly false. The variety of nature is just as marked as its uniformity; and, on the face of it, far more certain, since variety can be directly observed, whilst uniformity, strictly speaking, cannot. It is all very fine to adopt a haughty attitude towards particular propositions and to call them trivial; the fact remains that many such propositions are true, and that it is excessively difficult to state any principle which will favour laws as against particular propositions and not fly in the face of the facts. I can indeed state a principle of uniformity which will be compatible with any amount of variety, but I am far from sure whether it is either true or useful. The principle would be this:

$$\phi a.\psi a. \supset :(\exists \chi): \chi \neq \psi.\chi a: \phi x.\chi x. \supset_x.\psi x.$$

This means that if any individual a has the property ϕ and the property ψ [e.g., is a swan and is white] then there is some property χ other than whiteness [e.g., that of being European] which is possessed by a, and such that everything that is both ϕ and χ [e.g., is a European swan] is also ψ [e.g., is white]. The condition $\chi \neq \psi$ is added to avoid triviality, since if χ might be ψ a χ fulfilling the conditions always exists for $\phi x.\psi x$ analytically implies ψx. Of course χ might be identical with ϕ.

I am inclined to think that this is what those logicians like Prof. Bosanquet who say that all particular propositions are imperfectly apprehended universals have in mind. I am the more inclined to this view because this principle does make all laws simply convertible in a certain sense, and this is another characteristic opinion of the same school of logicians. Suppose that in the above formula we substitute everywhere ψ for ϕ and ϕ for ψ. We get

$$\psi a.\phi a. \supset :(\exists \chi): \chi \neq \phi.\chi a: \psi x.\chi x. \supset_x.\phi x.$$

Of course the χ will not in general be the same in the two cases; but it does at least follow from the principle that there is always an universal proposition with ψ as subject and ϕ as predicate as well as one with ϕ

19

as subject and ψ as predicate. And I can hardly suppose that these logicians intend to maintain much more than this.

Another principle, which many people seem to believe, can be deduced from the above. Many people would say that, if you find that some swans are white and that some are not, this is never the whole truth about the matter; all the white swans must have something common, and peculiar to them which 'accounts for' their whiteness.

A little simple logical manipulation leads to the proposition:

$$\phi a.\phi b.\psi a. \sim \psi b \supset : (\exists \chi, \theta):$$
$$\chi a.\theta b.\chi \neq \psi.\theta \neq \overline{\psi}: \phi x.\chi x. \supset_x. \sim \theta x.$$

e.g., If a and b are swans and a is white and b is not then there is another property χ possessed by a and a property θ possessed by b such that no swan with the property χ has the property θ.

4

Now the proposed principle, which we will call *Unax* for short, must be admitted to have certain merits. If Unax were true the problem of induction would be shifted and lightened. Without it we do not know whether there is any law connecting S with P; we are therefore liable to go wrong in two ways: (*a*) by thinking that there is a law and that we have discovered it when really there is no law at all, or (*b*) by thinking that the law is All S is P when really it is of some more complex form such as All SQ is P. If Unax be granted the first source of error vanishes. The second, which corresponds to the second difficulty in induction by simple enumeration, remains. But it could certainly be reduced by examining S's under as various conditions as possible. We could never end by being sure that the law took the simple form All S is P, but we might conclude with fair confidence that, *if* it be All SQ is P, the factor Q is pretty abstract and accompanies S under extremely variable conditions, so that for most practical purposes, it is negligible.

Unax also has the merit that it could never be refuted by experience. Whenever you seem to have a conjunction of attributes ϕ and ψ which is not an instance of a general law of the form $\phi x.\chi x. \supset_x.\psi x$ you can always say that this is because the property χ is too minute or obscure to be detected by our present means of observation. No one could refute this

THE RELATION BETWEEN INDUCTION AND PROBABILITY

possibility; and, if you believed it, it would furnish a motive for further and more accurate investigations.

This, however, is about all that can be said in favour of Unax. There remains much to be said against it. In fact Unax may be a first approximation to the principle for which we are looking; but it seems quite certain that, as it stands, it is in some ways far too general and in others not general enough, and that it is neither ultimate nor plausible. By developing these criticisms we may find out in what direction to look for more light.

(i) Unax, as stated, makes no difference between ϕ and ψ; they may be any properties or combinations of properties. Now when ϕ is a property like being a swan or a crow and ψ is a property like whiteness or blackness the principle seems plausible enough. But suppose that ϕ were a property such as being spherical. I hardly imagine that the statement that, if anything is spherical and white, then it possesses some other property χ, such that all spherical objects with the property χ are white, would seem plausible. It therefore looks as if ϕ and ψ must not be properties which are wholly unrestricted, and that in fact ϕ must be a property of a very special sort, if the statement is to seem plausible. This is reinforced by the following consideration. We have seen that, if we take Unax without any special hypothesis about ϕ and ψ, *two* laws correspond to every conjunction of attributes. Now many people would hold that if a swan is white there must be some property χ possessed by this swan such that all swans with this property are white. But how many people would hold that if a white object is a swan there is some property χ, other than that of being a swan, which is possessed by this white object and is such that all white objects with the property χ are swans? Yet this, as we have seen, equally follows from Unax, if ϕ and ψ are supposed to be subject to no special hypothesis in it.

(ii) For Unax a *single* conjunction of attributes is enough to make it certain that this conjunction is an instance of *some* general law. Nor is it easy to see how this could be otherwise, for the influence of *number* of instances seems to have been exerted in the only way in which it can be relevant, *viz.*, through the laws of probability, before ever Unax was invoked. I hardly see how any principle about nature which is to do the work required of it can refer to the number of observed instances. If it is about nature it is about what exists whether we observe or not, whilst the number of instances observed is at least partly dependent on our own actions. Yet many people who would agree that a good number of observed conjunctions of ϕ and ψ make it certain that ϕ and ψ are connected by a law would hesitate to say that a single such conjunction makes it even highly probable. It is important to be quite clear as to the precise nature of the difficulty here. (a) Nobody supposes that, with Unax or without, a single instance of ϕ conjoined with ψ makes the particular law that ϕ is always accompanied by ψ probable. But (b) Unax does say that a single instance makes it absolutely certain that there is *some* general law connecting ϕ with ψ. Now most people would be inclined to hold (c) that a fair number of instances of conjunction are needed to make even this probable, though a fair number will make it practically certain. Now their view is not supported at all by the probability-theory of induction *without* Unax; whilst, if they accept Unax as offered, their view is unintelligibly timid. Hence it must be supposed that they accept some principle about nature which is less sweeping than Unax; yet it is very difficult to see what principle about *nature* there could be which makes *number of observed conjunctions* relevant at just this point.

5

I am inclined to think that both these difficulties (i) and (ii) are to be met by the same modification. When do inductions by simple enumeration seem to be highly plausible and when not? They seem plausible when we are dealing with substances which are believed to belong to what Mill would call a Natural Kind. We believe pretty strongly in the results of such inductions when they deal with the properties of such things as crows or swans or pieces of silver. But no one attaches much weight to inductions about the colour of billiard balls or counters in a bag. If Unax is to be rendered plausible it must be subject to the restricting hypothesis that ϕ is a property or set of properties defining a kind. If this be granted we see why common sense will not allow the reversibility which Unax permits when ϕ and ψ are unrestricted. Unax now takes the form:

$$\phi \in K . \phi a . \psi a . \supset : (\exists \chi): \chi \neq \psi . \chi a : \phi x . \chi x . \supset_x . \psi x .$$

This we will call *Unaxk*. Now Unaxk says nothing about ψ defining a kind; hence, on substituting ψ for ϕ and ϕ for ψ, we get nothing startling, but merely a proposition with an hypothesis $\psi \in K$ which is in general false.

We can also see now why common sense wants a number of observed instances before it will consent to be sure that there is *some* law connecting ϕ with ψ. It wants these instances in order to persuade it of the truth of the hypothesis that ϕ defines a kind.

It can only feel sure of this when it has met with a fair number of instances of ϕ and found that they have a great number of properties beside ϕ common and peculiar to them.

Finally (iii) we can now admit that Unax is not ultimate, and can see why. Unax is only plausible in the modified form of Unaxk. Unaxk refers essentially to kinds, and we have not as yet analysed what is meant by kinds and what is involved in the assumption that there are kinds in nature. Any further progress in solving our problem will therefore depend on a careful discussion of this subject. We must therefore bid Unaxk a long farewell for the present and turn our attention to the assumption that there are natural kinds.

6

Even without entering at all deeply into the question of kinds we can see in a general way how the assumption of kinds affects the problem of induction about the properties of substances. Such inductions seem most plausible when the subject is a well-marked class like swans or crows and the predicate some fairly general and simple property like blackness or whiteness. Now the mere fact that ordinary language has taken the trouble to invent a general name like *swan* or *crow* tells us a good deal about nature. It implies that a large number of objects have been met with which have combined pretty constantly a large number of properties varying only within fairly narrow limits. It is true that you may *define* a crow or a swan or a man by a few properties. But this very fact is symptomatic. Whatever may be the dictionary meaning of 'man' *we* always mean by it something with a great many more properties than animality and rationality or twoleggedness and featherlessness. Anything that had these properties but differed widely in other respects from the men that we had met would only with great hesitation be called a man. Hence the fact that we are content with the dictionary definition is due to the fact that so far in our experience the properties mentioned therein have been associated with a whole bunch of other properties, and that all these have been exemplified together with but slight variations in a great number of instances. Thus when we ask ourselves the question: Are all S's also P? and suggest the possibility that some may not be P we imply that P is only one of a large number of attributes, and we imply that a slight variation in P is consistent with the bulk of the remaining attributes being unchanged. For with any large change, we should cease to go on calling the object *an S*, and thus, even if this object turned out not to be a P, this would not be relevant to the question whether all S's are P; for this object would not be counted as an S.

So the actual state of affairs in any induction about substances to which we should be inclined to attach much weight is this: (a) A large number of individuals have been observed all of which had a large number of attributes in common and only differed by small variations of these attributes within narrow and characteristic limits. Scarcely any individuals have been observed which agreed with the former in a great many respects, but otherwise differed profoundly from them. And if such have been observed and have been numerous they count as a different kind and have a different name, so that no question arises of treating them along with the former individuals in making our induction. (b) The attribute P has been found to be present in all these individuals. This attribute is not of such importance that a change in it alone would prevent an object otherwise agreeing with other S's from being called by the name *S*. (c) If there be other individuals which agree so far with those already observed as to be appropriately called by the same general name *S* as they, how probable is it that they will also agree in having the attribute P?

The superior plausibility of inductions about kinds is thus partly a matter of words; but, like most matters of words, it rests ultimately on a matter of fact. The purely verbal point is that, unless the unobserved objects resemble the observed S's in the vast majority of their attributes they will not be called *S's*, and the question whether they be P or not will be irrelevant to the question whether all S's are P. The factual basis of all this is that a large number of very similar individuals have been observed; if they had not been numerous and had not exemplified an outstanding bunch of attributes men would not have troubled to give them the special name *S*. Thus, in any actual induction, the evidence is never merely the *number* of examined instances, but also the predominant agreement of all these instances with each other and the presupposition that the doubtful and unexamined cases must predominantly agree with the examined ones in order to count as relevant instances for or against the suggested law.

We might put the argument in the following way. The objections to induction by simple enumeration about the properties of substances are unfair to that process in the only case where anyone attaches much weight to it. They are unfair for two reasons: (a) They do not state the problem properly; and (b) they do not consider the whole of the evidence. Let us consider these points.

(a) It is unfair to put the problem in the following form: 'All the observed S's are P. There are innumerable unobserved S's. What is the probability on your observation that all these are P?' For what is the evidence that there are innumerable unobserved S's? Surely it is of just the same kind as the alleged evidence that the unobserved S's are P. You have observed a large number of S's; they were all P. If the observation of a large number of observed S's be a good ground for thinking that there are innumerable unobserved S's it would seem to be an equally good ground for thinking that there are innumerable SP's; for all the observed S's were in fact SP's. I do not at present wish to assert that we have good evidence for *either* conclusion; but it is obviously unfair to talk as if we were certain of the former and to make this a ground for feeling doubtful about the latter. It does seem likely that anything that is evidence for the one will be in its degree evidence for the other. We might put the matter thus. Either your evidence makes it highly probable that there are unexamined S's or not. If so, it is difficult to see what evidence could make it highly probable that there are unexamined S's and leave it highly improbable that they are SP's, when all the examined S's were SP's. If, on the other hand, there is no strong reason to believe that there are many unexamined S's, there is no strong reason for putting the probability that all S's are P very low, for there is no good reason to think that m is very small as compared with n in the fraction $\frac{m+1}{n+1}$. (It must be understood that at present I am only using general arguments, which must be taken as *illustrating* the way in which the assumption of kinds might affect the theory of induction, and not as *proving* anything conclusively. We shall have to consider the whole question in much greater detail when we have learnt more about kinds.)

(b) To consider only the number of the observed S's is to neglect part of the evidence. We have also to remember that to be called an *S* at all an unobserved object has to resemble in most of its properties those objects which were observed and were P. Hence an argument by simple enumeration is always also an argument by analogy, and, *ex hypothesi*, the analogy is very strong or the unobserved case does not count as an instance for or against the law about S's.

7

We see then that any actual induction about the properties of substances involves at least two presuppositions beside the numerical and other data of the argument, *viz.* (a) that we are dealing with *substances* and (b) that there are *natural kinds* of substances. Anything that is involved in these two assumptions may therefore fairly be regarded as part of the actual premises or principles of such inductions. We must therefore see what these two assumptions really do amount to, and afterwards what evidence there is for them. We shall find that, as regards evidence, (a) and (b) are entangled with each other and with induction by simple enumeration in a highly complicated way. But we must begin by treating them separately.

a. *The assumption of substances*

When we call a swan a substance we imply that it is something that persists at least for a time; is distinguishable from other swans and from other things which coexist with it; and that, in spite of changes, we can in theory at least identify it as it is at one moment with itself as it was at other moments. A persistent, changeable, and yet identifiable substance is always *at least* a series of states having certain relations to each other and certain properties common to them all. It may be something more than this, but I do not think that it need be so. By a state of a thing I mean a momentary particular which is one of the whole series of related particulars constituting the thing. A state is thus a 'substance' in the logical sense of being a particular and not a universal, though not in the physical sense which involves persistence and identity through change. When I call these states 'momentary' I do not wish to tie myself down either to the view that they have no duration or to the other view that each lasts for a very short time, characteristic perhaps of the series in which they occur. For our present purpose the difference is not of much importance. When I say that θ is a state of the substance Θ I therefore mean that θ is a particular which is momentary in a loose sense and is one of a series of momentary particulars $\theta_1, \theta_2 \ldots$ which have the sort of common properties and mutual relations which entitle such a series to be called a substance. (This view is to be distinguished from the assertion that 'things are *classes* of their states'; it says that things are *complexes* of their states and complexes of a very special kind. To illustrate by an analogy: My face is a complex in which my features are elements; it is not the class of my features.)

To say that Θ persists up to the time t means that there are θ's fulfilling those conditions up to that time. To say that it then ceases to exist means that after then there are no θ's which have the right amount in common or the right kind of relations with those of the series $\theta_1, \theta_2 \ldots$ which existed before t and were the states of Θ. To say that Θ persists but changes at t means that there are θ's which exist after t and have enough similarity to and continuity with those which exist before t to be counted as states of the same thing Θ, but that the last to be observed of the latter and the first to be observed of the former differ from each other in some 'first-order property'. By a 'first-order property' I mean a singular proposition ascribing a 'lowest quality' to a definite particular state, or asserting a 'lowest relation' between two or more definite states. I use the phrases 'lowest quality' and 'lowest relation' by analogy to the phrase *infima species*. I should not call colour, or even red, a lowest quality, but only a perfectly definite shade of red with definite intensity and saturation. In fact a lowest quality is universal in that it can have a plurality of instances; but these instances must be particulars. Similar explanations apply to the phrase 'lowest relation'.

The next point to notice is that all properties of *things* are at least 'second-order properties'. By a 'second-order property' I mean the assertion that a propositional function whose particular values are first-order properties gives true propositions for all, some, or certain values of the variable. Now it is evident that a great many properties of things are assertions about their characteristic ways of behaving. They thus assert how the first-order properties of one state will differ from those of an earlier state under given circumstances. Evidently such assertions are at least second-order properties. But this is equally true about what are called 'permanent properties' of things, though the fact is here less obvious. When you say that this penny retains its mass through all physical and chemical changes you are saying that for all values of θ, such that θ belongs to the series of states Θ constituting this penny, the function 'θ has the mass m' gives a true proposition. The permanence of an attribute is thus only a rather special and peculiar mode of behaviour, and the persistent properties of substances are of at least the second order just as much as assertions about their characteristic ways of changing.

8

Doubtless permanence in this sense is the earliest and most striking feature which is chosen as a criterion to judge whether a state belongs to a series constituting a thing. Many series do continue in our experience for long periods with scarcely any serious variation in their first-order properties from one state to another. But even such series, which uneducated common sense regards without hesitation as constituting persistent things, have long gaps as far as our experience is concerned. While our attention is otherwise occupied those series may continue, but we certainly have no direct evidence that they do. How does common sense fill in such gaps? Suppose we are aware of a series of very similar states which we regard as the thing Θ_1; suppose that there is then a gap in our experience and that we then meet with no more states of this kind for a time. Lastly suppose that we again meet with a series which we can regard as a thing Θ_2, and that the states of Θ_2 are as similar to those of Θ_1 as those of Θ_1 are to each other. Under what circumstances do we regard Θ_1 and Θ_2 as the same thing? (a) We may find that whenever we choose to adjust our bodies as they were adjusted when we perceived Θ_1 we are aware of a state θ as like those of Θ_1 as the latter are to each other. Under these circumstances we should say that Θ_1 persisted and was the same as Θ_2. (b) On the contrary we may of course find that a change of bodily adjustment is needed in order to perceive Θ_2, and that we can only become aware of a θ whenever we choose, provided we suitably alter the adjustment of our bodies. In such cases we tend most strongly to identify Θ_2 with Θ_1 and to hold that Θ_1 has really persisted through the gap in our experience, provided that we find that in order to become aware of θ's intermediate between the end of Θ_1 and the beginning of Θ_2 an intermediate amount of adjustment is needed between that which was required to be aware of the last θ in Θ_1 and that required to be aware of the first θ in Θ_2. The point here then is that you can perceive a θ of the right sort at any point in the gap if you will make the right bodily adjustments, and that the right bodily adjustments for success at various points in the gap form a continuous series between those which are successful at the beginning and those which are successful at the end.

We thus see that an important criterion for the persistence of a thing Θ is the belief that *whenever* we choose to perform certain actions we shall observe a particular θ which is so connected with the θ's that actually are observed as to count as a state of the same thing. Now what evidence can I have for this belief in the case of some definite thing Θ which has ceased to be under my observation for a certain ten minutes? Clearly I cannot know by direct observation of Θ that if I do the right things in the ten minutes' interval I shall perceive a θ which can be taken as a state of *it*. For, by hypothesis, I do not do the right things, and do not become aware of any such states within this interval; this is implied by saying that Θ ceases to be under my observation during that ten minutes. My only evidence (apart from the testimony of others, which is often lacking) is the behaviour of *other* things of the *same kind* as Θ on other occasions. Suppose, *e.g.*, that I observed a certain state θ_1 at the beginning of the ten minutes, and that at the end of it I began to observe a certain state θ_2. By hypothesis I have observed no intermediate states of this particular Θ. But I may have observed other Θ's at other times. I may have observed one of them for two minutes after it reached a state like θ_1, another for five minutes, another for seven, and so on. I may even have observed a Θ for a complete ten minutes after it attained a state like θ_1 and I may have found that it then reached a state like θ_2. Thus my evidence for supposing that at a given moment in an interval during which Θ was not under observation I should have observed a certain state θ_m if I had done certain things is that I or others actually have observed a state like θ_m at a corresponding period in the history of some other Θ which was under observation.

We thus see that the logical relations between substances, natural kinds, and induction are extremely complex. (i) Obviously the assumption of kinds of substances involves the assumption of substances. But (ii) we should have very little evidence for the persistence of a given substance if it were not for the fact that other substances of the *same kind* are observable when it ceases to be under observation. (iii) Inductions about the properties of substances are not plausible unless those substances are supposed to belong to a natural kind. Yet (iv) the evidence for the persistence of an unobserved substance from that of others of the same kind is itself inductive. (I do not of course suggest for a moment that people actually reach the belief that their table continues to exist when everyone goes out of the room by inductive arguments from the behaviour of observed tables. They do not *reach* such beliefs by argument at all, any more than they argue to the existence of physical objects from their sense-data or to that of other minds from the behaviour of other bodies. But, if their belief in the persistence of a given substance were challenged, the only grounds that they could offer would be inductive arguments from other substances of the same kind which had remained under observation.)

It will now be wise to discuss the assumption of kinds, since we see that it is closely connected with the persistence of substances and it is part of the definition of a substance to be a more or less persistent series of states.

9

b. *Assumption of kinds*

If we consider all the momentary states of all the material things which we have met, we find that, though infinitely various, they ring the changes on a comparatively few variables. States differ from each other in colour, sound, taste, smell, temperature, shape, size, etc. But they agree in being determined by one or more of these variables and by some special values of them. Let us call the various sensible qualities – colour, sound, temperature, 'feel', smell, taste, etc. – *primary variables*. The above list is practically exhaustive as far as human beings are concerned. I have excluded shape and size from the list for reasons which will appear in a moment. Each of these primary variables has a comparatively small number of *dimensions*, as I will call them. *E.g.*, the dimensions of sound are pitch, loudness, and quality. Dimensions are specifications of a primary variable, having the following properties: (i) In any definite instance one value of each dimension must be specified; (ii) *A priori* and apart from any special causal laws which may be found to hold in this particular world any value of one dimension may coexist with any value of any other dimension of the same primary variable. Lastly each dimension of each primary variable is susceptible of a range of possible values which is sensibly continuous.

The position of spatial properties is unique and peculiar. We cannot treat shape and size as themselves dimensions, for they cut across the primary variables; *e.g.*, a patch of colour and a patch of temperature both have shape and size. On the other hand we cannot treat shape and size as primary variables. For it is of the essence of primary variables to be antecedently independent of each other. There is, *e.g.*, no synthetic, *a priori* proposition asserting that colour must be accompanied by temperature or temperature by 'feel' (in the sense of hardness or softness), even though some such propositions should be found to be true in the actual world. Now there are *a priori* connexions between spatial attributes and primary variables. All instances of colour and temperature and 'feel' at least have some shape and size. And all instances of shape and size are also instances of *some* primary variable, *e.g.*, colour or temperature or 'feel'. We may say then that as regards any given primary variable extension behaves like a dimension, *i.e.*, it must be specified to determine any particular instance. But, unlike a genuine dimension, it is not tied down to any one primary variable. Finally extension in itself of course has dimensions in the strict sense.

Now any momentary state is completely specified when we are given (a) the primary variables, (b) the values of each dimension of each variable, and (c) the extension of the determinate value of each primary variable. The sum total of all antecedently possible combinations of values of this kind would give all the antecedently possible sorts of states at a moment. Any one of these sorts of states might, so far as we can see, have any number of instances. The only antecedent restriction on the number is that two precisely similar states will not count as distinct if they completely overlap each other in space. Now antecedently there seems no reason why any one of the possible sorts of states should be represented in nature by more instances than any other. We might therefore have reasonably expected to find at any moment the whole multiply-continuous series of possible sorts of states about equally represented in the existent world. But our actual experience of the world has been utterly and flagrantly contrary to this expectation. What we have found is not a regular distribution of all the states at a moment among all the possible sorts of states, but a 'bunching together' of instances in the neighbourhood of certain sorts of states. Intermediate possible sorts are scarcely represented in nature, so far as our experience has gone, at all.

Suppose, *e.g.*, that there are N primary variables. Then of course there are ^NC_r possible r-fold combinations of them, and the total number of combinations of all orders will be $2^N - 1$. Now let us confine our attention to any one of the ^NC_r r-fold combinations of primary variables. Each of the r variables will have a finite number of dimensions, and between them they will possess a number of dimensions which may be represented by pr, where p is a positive integer in general greater than 1. Imagine now a pr-dimensional space formed with one dimension of one of the r variables for each of its axes. Then, setting aside the characteristics of shape and size which, as we have seen, are also needed completely to specify a possible sort of state, we may say that each point in this space represents a possible sort of state defined by this particular selection of r out of the N primary variables. Now suppose that a fluid were distributed throughout this space in such a way that its density at any point represents the number of instances in the world of the sort of state represented by the point. Let us further suppose that the density of the fluid at a point were represented by the blackness of a dot made at that point. Then antecedently to experience we might expect this space to be uniformly shaded. But in actual fact, so far as our experience has gone, we have found a quite different arrangement. We should find a number of blobs in the space surrounding certain points. These blobs would be very dark near their centres and would shade off very quickly in all directions as we moved away from these centres. In the regions between the blobs there would be practically no dots at all, and such as there were would be extremely faint. And lastly the whole set of blobs would be confined within a region defined by moderate values of the variables.

10

This sort of distribution corresponds to what is meant by natural kinds. A natural kind is a region containing a blob. To drop metaphors, a natural kind of state is a sort which has a predominantly large number of instances in nature and such that the number of instances of neighbouring sorts of states falls away quickly in every direction. The sort which has the maximum number of instances (and in our spatial picture is the mean point and the blackest of a blob) is the *type* of the kind in question. Any particular instance of it or of its adjacent sorts counts as a state of the kind. A kind of substance is, to a first approximation, a series of states all of a kind, and possessed of the sort of continuity and relations which make them one substance. (I say to a first approximation, because, as we shall see later, characteristic modes of change are as typical of kinds of substances as constancy of kind throughout a series of states.)

The net result then is that, even to a superficial observer, the distribution of states at a given moment is about as far removed as it could be from what is antecedently most probable, and that this mode of distribution shows no sign of becoming more uniform when we take all the moments of human experience together.

Now either this habit of heaping instances round a comparatively few possible states is typical of nature as a whole or it is not. If it is not we have to explain as best we can why it has been characteristic of nature so far as it has come under the notice of human beings. Supposing, for the sake of argument, that nature as a whole really distributes its instances uniformly among possible sorts we shall have to go on to assume that the position of the human race is in some way wildly abnormal so that the parts of nature which have fallen under its observation have been utterly non-typical of the whole. What would this assumption amount to?

It might mean either that the human race had been confined to a section of the universe in which the distribution of instances is excessively unlike their distribution over nature as a whole, and that this exaggeration in our part of the universe is corrected by complementary exaggerations in other parts. Or it might mean that, even within the part that has fallen under our observation, the distribution of instances is really pretty uniform, but that limitations in our perceptive powers or in our interests have prevented us from noticing all but the instances of a few possible sorts. In the end both alternatives depend on supposed limitations of our powers of perception. The second explicitly does so. The first, on further consideration, is easily seen to do likewise. The only importance of space and time for the inductive problem is that they impose limitations on what we can directly observe, and hence at the same time provide the motives and limit the data for inductive arguments. I cannot directly observe what is very remote in space or what happened before I was born, nor can I now directly observe anything that is going to happen later unless I chance to be a prophet.

11

Now the lack of uniformity in the distribution of instances *within* the region to which I have been confined by spatio-temporal limitations certainly cannot be explained wholly by limitations of my interests and powers of perception. No doubt if the values of primary variables be above or below certain limits I cannot observe them. No doubt, too, there may be many variables that cannot fall under my observation because I lack the needful sense-organs. But this will not account for my failing to observe instances of sorts which fall between the sorts of which I do observe instances. The fact that I occasionally do observe instances of these sorts (*viz.*, 'monsters' in an extended sense of the word) shows that their rarity in my experience cannot be explained by supposing that they are really present in large numbers but are unobservable to me. Again, while it is true that I often slur over minor differences and treat instances as exactly alike when they are only rather similar, it is certainly not true that my interest is only excited by similarity and not by difference. The success of Messrs. Barnum and Bailey shows that it is not mere lack of interest for intermediate sorts that makes us ignore them. If, *e.g.*, pig-faced ladies were not *really* rare within the range of our physically possible experience it would be unintelligible why the few who do turn up should be so much more interesting than ladies of the more usual kind. Thus I think we are forced to conclude that that part of nature which falls within the spatio-temporal limits of possible observation really departs very far from a uniform distribution of instances among possible sorts; and that the appearance of departure from uniformity cannot be explained by limitations of our interests or powers of observation.

12

The second alternative, that the part of the world that has fallen under human observation really does depart widely from uniform distribution but that this is averaged out by the much wider part that has never been

THE RELATION BETWEEN INDUCTION AND PROBABILITY

observed, is much harder to treat properly. It evidently assumes that there is an unobservable part of nature and that the sole reason why it is unobservable is because we cannot perceive what is very distant in space or past time or what is future in time. This assumption itself has doubtless many implications, but for the moment we will take it as it stands. We may then represent the whole course of nature as contained in a four-dimensional space with three spatial and one temporal axis. We may regard a human observer as a point surrounded by a four-dimensional solid. This solid represents the spatio-temporal limits of his possible perceptions. The human race within historical times will be represented by a big four-dimensional solid composed of such solids. Of course the solids will not exclude each other wholly; the centres of one or more will often lie within those of another. Thus the solid will be rather like a mass of bubbles made by blowing through a pipe into soapy water. The limits of this solid will be those of possible human observations within the period for which human history has lasted. Now either (a) we may neglect the fact that the human race arose from definite causes in a definite part of the universe, or (b) we may take it into consideration. Let us first neglect it.

Then antecedently we can regard this solid representing possible human experience as shot at random into the space representing the whole course of the universe, *i.e.*, we have no ground antecedently for thinking that it is more likely to fall in one part of the course of nature than in any other part of the same shape and duration. The actual content of human experience will be represented by the content of the part of the whole four-dimensional space into which the four-dimensional solid happens to fall. Now if the heaping of individuals about kinds be a peculiarity of a small section of the universe, whilst elsewhere the distribution is nearly uniform, it is highly unlikely that human observers will have happened to fall just into this part of the universe. The larger we suppose the universe to be compared with the part of it which has this peculiarity the less likely it is antecedently that the solid representing the limits of human experience should have fallen totally inside this peculiar region. Really we have three four-dimensional volumes to compare: (a) that representing the whole course of nature, (b) that of the solid representing the spatio-temporal limits of historical human observation, and (c) that of the supposed exceptional region within which a discontinuous distribution of individuals about a few natural kinds is supposed to hold. Unless (c) be very small compared with (a) we cannot be very far wrong in extending the characteristics of what we have observed to the whole universe. On the other hand if (c) be very small compared with (a) it is very unlikely that (b) when thrown at random into (a) should fall wholly inside (c). And it is obviously more and more unlikely the nearer (b) approaches in volume to (c). Now it is only if the general course of nature changes soon after the spatio-temporal limits of our present experience are surpassed that the inductive extension of the general characteristics of what we have observed will soon lead us wrong. That is, such an inductive extension will be *practically* harmless *unless* (b) nearly approaches in volume to (c); and we have just seen that *if* (b) nearly approaches (c) the fact that (b) has wholly fallen inside (c) is an extraordinary coincidence which renders the existence of the supposed exceptional region (c) highly improbable.

13

But it will no doubt be objected at once that all this talk about the human race being 'shot at random' into the universe like a sack of coals into a cellar is the merest nonsense. It actually did arise at a certain moment in certain parts of space where the right conditions were fulfilled and has gone on ever since. Hence its range of experience cannot be compared to a *movable* solid which might have fallen anywhere in the universe. Now these statements may very well be true – I suppose that we all believe that they are true – but are they relevant? What is a person who makes them assuming? He is assuming that he can write a hypothetical history of the origin of human observers. Now this means that he supposes himself to know (a) that certain conditions held before human observation began, and (b) that these conditions, operating according to certain laws, were necessary (if not sufficient) for the production and continuance of life and mind as we know them. He thus claims a knowledge of what existed outside the range of human observation and of the laws that it obeys. His only ground for this must be the belief that he is justified in extending the characteristics of the part of the world that has fallen under human observation to parts of it which, by hypothesis, cannot have done so.

The logical position therefore seems to be this. Either we know that the general characteristics of nature which we have observed (confinement of instances to kinds, regularities of behaviour, etc.), are equally characteristic of the parts of nature which we have not observed or not. If so, then it is doubtless nonsense to talk of the human race and its observations being as likely to fall in one part of the total course of nature as in another, and our previous argument will be useless. But then it will also be needless. For anyone who supposes himself to have this knowledge supposes himself to know that the part of nature that has fallen under observation is not peculiar in its general (and even in some of its more special) characteristics. If, on the other hand, we entertain a doubt whether the general characteristics of the observed part of nature hold of the unobserved parts we *ipso facto* leave open the possibility that these unobserved parts are subject to no special laws and do not confine instances to kinds. Now relative to that possibility it is not nonsense to talk of the actual position of the human race in the course of nature as a whole as a random position. And what we have argued is that the hypothesis that we are in a singular region of nature tends to undermine itself because it is highly improbable that the whole course of human experience should fall (as it has done) into what on the hypothesis itself is a small exceptional region of the universe. It must be noticed that this argument only applies at all strongly to the *general* characteristics observable in the part of the universe that has fallen under observation. It would be very extraordinary that, if only a small part of the course of nature confined its instances to kinds and its changes to regular rules whilst the rest of it did nothing of the sort, human experience should have happened to fall wholly within that small region. But it would not be at all extraordinary if in other parts of nature certain kinds which are predominant with us are not represented and conversely. In fact it is obvious that our experience makes it much more probable that the *general* characteristic of confinement to kinds extends widely beyond its limits than that the more *special* characteristic of favouring such and such kinds is widely extended. For the more special proposition implies the more general and not conversely; so that whatever is in favour of the former is in favour of the latter, but there may be evidence for the latter which has no special relevance to the former.

14

Extension of theory of kinds

So far we have argued that, even to a superficial observer, nature appears not to distribute its instances equally among possible sorts, and that it is reasonable to regard this general characteristic as probably extending much beyond the limits of human experience. But, to a superficial observer, confinement to kinds, though a striking characteristic of the observed part of nature, is by no means an universal rule within this part. In the first place there are occasional 'monsters'. Then again the contemporary states of various substances which would be counted as of one kind are never exactly alike. *E.g.*, the swans or crows that exist at any moment all differ more or less in their first-order properties. Again, if instead of thus taking a cross-section at a given moment, we consider the series of states constituting a given substance, they differ from each other in many first-order properties. And a point may be reached at which either the series stops altogether and the substance is said to have ceased; or else the first-order properties may change so radically whilst certain conditions of spatio-temporal continuity are still fulfilled that the substance is said to have 'changed into' one of another *kind*. There can be no doubt, I think, that the face of nature does present these aspects to all of us whilst we are still 'trailing clouds of glory behind us', and that it continues to do so to many until the end of our lives.

Now at this stage there enters a characteristic habit of the human mind which has constantly operated with highly useful effects in the history of science. We draw a distinction between the superficial appearances of things and their more detailed and latent character. A contemplation even of the superficial aspects has strongly suggested to us some general rule, but there are a certain number of apparent exceptions. We then tend to proceed on the assumption that this general rule really is true without exception when the latent parts of nature are taken into consideration, and that the apparent exceptions can be explained compatibly with this view. Then we make more careful investigations with this idea as our guide, and we find that in a great number of cases the more accurately analysed and observed facts support the assumption. If this be so we tend finally to take the rule as a principle and to assume that any small residuum of obstinate facts which apparently refuse to come under it only *appear* exceptional because we have so far failed to find the right way of analysing or observing them.

I imagine that this is what M. Poincaré had in mind when he talked of laws being raised to the rank of 'principles' and then being 'true by convention' and 'beyond the attacks of experience'. It is important for us to consider the logical position of this habit. (i) In the first place we suppose that the law is strongly suggested to us by superficial observation. Now the law that all things are instances of kinds is quite as strongly suggested to us by observation as (say) the law that bodies continue to move uniformly in straight lines except for the action of other bodies. (ii) Our everyday experience has given us every reason to draw a distinction between things as they appear at first sight and things as they appear on closer inspection. Since things exhibit fresh details to us the more closely we observe them it is perfectly reasonable to suppose that they contain parts and details that we cannot observe at all. And, since the details that closer observation reveals are often found to be more important than those which were observable on a more superficial view, it is not unreasonable to think that the details which cannot be directly observed at all may be more important than any that can be observed. (iii) We have plenty of experience both of substances coalescing and of their separating; we know that the coalescence of two substances of the same kind generally gives a substance of that kind; that the coalescence of two of a different kind often gives one with different characteristic properties from either; and that sometimes when a substance splits up it does so into several of the same kind as itself and sometimes into substances of different kinds. Now all these facts, which are common enough when we examine the world at all carefully, help to make the theory of kinds, which is so strongly suggested, but not wholly confirmed by superficial experience, more and more definite and rigid.

The notion of compounds and mixtures which differ markedly in their superficial properties from their components is suggested by experience of actually mixing and separating substances. Once suggested and recognised as a fact in the region of nature with which we have dealt, it enables us to hold that those things which are not on the face of them instances of kinds may yet be mixtures or compounds of things which are genuine instances of kinds. Thus one exception to a rigid theory of kinds (*viz.*, the existence of things of intermediate sorts) is removed by following out a suggestion which is (a) made plausible by our experience so far as it has gone, and (b) which that experience in its gradual development suggests to be extensible beyond the limit reached at any given moment by actual observation. But we cannot stop here, for we are still left with the fact that contemporary instances of the same kind that have actually fallen under our observation are not exactly alike, and that the successive states of what we regard as a single substance of a kind may differ seriously from each other. It is in connexion with these problems, I am inclined to think, that the notion of causation and of conditions becomes prominent.

15

Kinds, substances, and causation

We here meet again that irritating interweaving of various fundamental notions which we have already had occasion to notice and which makes it so difficult to treat the subject in any satisfactory logical order. Causal laws refer to the states of substances and special causal laws to the behaviour of special kinds of substances. But on the other hand, as we shall see, the definition of a kind of substances partly depends on the causal laws which substances of the kind are supposed to obey. And the identity of a substance of a kind may itself be defined by the fact that the states possesses certain properties which figure in some special way in a causal law. Let me illustrate before going further. Silver is a kind of substance, and the superficial marks of the kind are certain physical properties like colour, hardness, specific gravity, etc. Yet the vast majority of the silver in the world at any moment is not represented by states with any of these properties; since most of it exists in chemical compounds of various sorts. A chemist in stating what he meant by silver would hardly trouble to mention these first-order properties. What he would do would be to mention how silver reacts under various conditions with various other substances. And he would count the characteristic properties of the various compounds of silver as much more distinctly characteristic of silver than the superficial properties of the metal itself. Thus when he talks of the characteristic properties of the kind of substance called *silver* he scorns to give us a mere enumeration of first-order properties, because he knows that these are constantly changing and that if he confined himself to them it would hardly be plausible to count silver as a kind at all. Instead he gives us second or higher-order properties, *i.e.*, statements of the characteristic mode of variation of the first-order properties under given conditions. Thus the characteristic marks of a kind involve conditions and causation. On the other hand all these higher-order properties themselves involve a reference to kinds of substances. They include statements as to what silver does in presence of chlorine, in presence of sulphur, and so on. Yet again these other kinds are themselves mainly recognised and defined by what substances belonging to them do in presence of other kinds of substances. If it is part of the 'definition' of silver that it is the kind of substance which gives a white insoluble compound with chlorine, it is equally part of the 'definition' of chlorine that it is the kind of substance that gives a white insoluble compound with silver. Lastly, when the chemist states all these second-order properties of silver he does not profess to be announcing merely analytical propositions; they cannot therefore be part of the meaning of *silver*, which must therefore be assumed to be known before the propositions are asserted. How are all these tangles and apparent circles to be straightened out?

I take it that the solution is somewhat as follows. The notion of silver as a kind of substance was first suggested by bits of metallic silver seen and touched under certain 'normal' conditions of illumination, etc. These first-order properties continued much the same through long series of states which had the sort of continuity with each other that constitutes them states of one thing. They were taken as the original definition of silver. But silver, defined in this way, is continually ceasing to exist as circumstances change. It is found however that when a 'silver series' stops and is replaced (say) by a 'silver chloride series' certain regularities of mass hold between the two series, and under suitable conditions the 'silver chloride series' can be stopped and replaced once more by a 'silver series' in the old sense of silver. The mass of each state of this second silver series is the same as that of the first silver series. This identity of mass and of other first-order properties, the spatio-temporal continuity of the two silver series by the intermediation of the silver chloride series, and the regularity with which the silver series passes into a silver chloride series under one set of conditions and conversely under another, enable us to identify the first silver and the second. And these facts are summed up in the statement that the silver continued to exist throughout the silver chloride series in spite of appearances to the contrary. Now regularities of precisely the same kind hold for sulphur, chlorine, etc., defined originally by certain superficial first-order properties which persist under 'normal conditions'.

INDUCTION, PROBABILITY, AND CAUSATION

16

We thus arrive at a distinction of kinds into kinds of the first, of the second, and (as we shall see in a moment) of higher orders. Kinds of the second order (chemical compounds) are true kinds in the sense in which we have all along been using the word. But the instances of them begin and cease in the course of history. This always happens, so far as our experience goes, by the coming together or separation of instances of kinds of the first order (chemical elements). Instances of kinds of the first order are taken to be persistent and not to have begun or ceased in the course of human experience. And this view is held in spite of the fact that such instances are constantly disappearing and apparently coming to an end; for, after all, chemical elements are much less common and less stable than chemical compounds. The explanation of this apparent paradox is however quite simple after what has been said above. The kinds which are so noticeable even on the most superficial view of the world are mostly of the second or third order. Swans, crows, etc., are kinds of the third order; for they consist of instances of certain kinds of the second order in certain characteristic proportions, arrangements, and extensions, about which they vary within narrow limits. The main reason why these are the kinds that strike us is their comparative stability. By this I mean that each instance of such kinds consists of a series of states with first-order properties which vary very little even though conditions change a good deal. This is of course less true of kinds of the third order than of many of the second, for crows and swans die and decay, but many chemical compounds are intensely stable towards quite enormous changes in conditions. We can see then why it is kinds of the higher orders which first attract our attention and suggest to us the notion that confinement of instances to kinds is a general characteristic of nature, and that if we look more carefully we shall find that it is a rigidly general rule in spite of superficial appearances to the contrary. But, when we do investigate more closely, we find that these kinds which first struck our attention are not as a rule the most important kinds in nature. *E.g.*, silver chloride, as defined by its common physical properties, is an extremely stable kind; *i.e.*, these properties persist through long series of states under highly variable conditions. Compared with it silver, as defined by its common physical properties, is an unstable kind, for it is constantly tarnishing, dissolving, reacting, and so on. But under certain conditions a silver chloride series does wholly change its first-order properties and is succeeded by a silver and a chlorine series. Now we have no ground for saying that the silver chloride really persists after the change; for, if it does, does it do so in the silver series or in the chlorine series? It seems arbitrary to choose either. Again the mass of the silver chloride is now divided between the two series, and no silver chloride can be got from any one of them till either the other itself or an equal mass of some different sample of it is added to the first. We thus can attach a definite meaning to the statement that bits of silver and masses of chlorine persist in spite of appearances to the contrary; but, when we define persistence in this way, we have to deny that a bit of silver chloride persists when a silver chloride series ceases to show its defining first-order properties. Thus we reach the notion of first-order kinds and see that they are more important though less obvious superficially than those of higher orders.

At this stage the extremely peculiar character of the part of nature that has fallen within human experience becomes still more marked. For we find that every bit of matter that we come across can be regarded as either an instance of a kind of some order or as a mixture of instances of various kinds, and that the number of distinct first-order kinds is ridiculously small. We admit of course that there may be first-order kinds that we have never met with, and that what we take to be a first-order kind, may prove to be of a higher order. But we do seem to have hit on the general groundplan of the material world, however inadequate may be our knowledge of the details. And that ground-plan, suggested to us even by a superficial observation of nature, has shown itself to be capable of statement in a more and more rigid and exacting form as we have investigated nature more and more carefully.

17

We have now seen that many of the most interesting properties of kinds of substances are not assertions about the *persistence* of the first-order properties of states of a series, but assertions about the ways in which such properties vary from state to state of a series with varying conditions. However Irish it may sound, it is true to say that the most important properties of first-order kinds are properties of second-order kinds. This of course simply means that, *e.g.*, the most important properties of silver are not the superficial physical properties of metallic silver, but are statements of the conditions under which metallic silver turns into such and such compounds and the conditions under which such and such compounds again give metallic silver. Now the identification of 'such and such' a compound of silver (*e.g.*, silver chloride) can only be made by mentioning enough of its properties to characterise it unambiguously. Thus it is true that most statements about first-order kinds are statements about the properties of the second-order kinds into and out of which they pass under given conditions.

Again, it is probably true that we should not have troubled much about conditions if it had not been for the changes in first-order properties that occur along a series of states regarded as constituting a thing. If first-order properties had all been highly persistent with varying conditions we should probably not have noticed that they depend on conditions at all. But, as it is, the variations in many series of states having thinghood force the notion of conditions on our attention, and then we come to see that even persistence of first-order properties depends on conditions and is only relative. Change the conditions enough and the most persistent first-order properties will begin to vary.

Now I am inclined to think that the notion of causation and conditions is best regarded as an attempt to reconstruct at a higher level the crude notion of things which has broken down on reflexion and minuter observation. I think that we shall see this clearly if we consider what is commonly believed in practice about causal laws and the Law of Causation. In the first place it is always changes that are felt to need explanation, *i.e.*, if the series of states constituting a thing varies from state to state in first-order properties we are not inclined to accept this as an ultimate fact. Parallel with this, but less often explicitly noticed, is another fact. We find instances of the same kind coexisting at different places in space. Though we count them of the same kind the contemporary states of several of them will not as a rule be exactly alike. All crows are instances of a kind, but at every moment there are small differences between one crow and another. This is felt to demand some explanation. The cause of demands such as this should now be fairly obvious. Our original criterion of the persistence of a given thing was identity of first-order properties throughout a series of states possessed of spatio-temporal continuity with each other. In so far as the first-order properties vary throughout such a series the series departs from the standard of a persistent thing. Hence the need of an explanation for *changes* and the absence of need for an explanation of *persistence* is the need to reconcile a contradiction. We are determined (a) to go on talking of *this* thing and saying that *it* persists; indeed this is implied by calling the change a change in *it*. But (b) our original criterion of identity uses persistence of first-order properties. The need for explanation of change is the need for a less simple-minded criterion of one thing and of the persistence of a thing, which shall be compatible with both change and identity.

Again our ideal kind, suggested to us but never wholly exemplified in the world as we have found it, would have a large number of *exactly* similar instances. Actually we find large numbers of *very* similar but partly different states coexisting in various parts of space. Our demand for explanation is the demand to be allowed in some way to keep our notion of kinds as possessing *exactly* similar instances and yet to admit that the contemporary instances very rarely are exactly alike.

18

These two closely connected demands are, I think, to be regarded as being in the strictest sense postulates and not axioms. They set us a problem, but there is no guarantee *a priori* that it will be soluble. What I mean is that it is not in the least self-evident that the universe *must* respond to our demand for permanent substances and for ideal kinds in some new sense of permanence and of kind, when it has failed to answer completely to our original criterion. The actual fact seems to be this. The world as it presents itself to superficial observation fulfils to a highly surprising extent the condition of consisting of permanent substances of a few marked kinds. It fulfils this still better when we investigate more closely. But it does not fulfil it altogether. The position is that it fulfils it so well as to raise the expectation that a modification of the definition of permanence and of kinds, which shall be in the spirit of the original definitions, can be found, and that with this definition the universe will *strictly* consist of permanent substances belonging to a few ideal kinds. I am prepared to believe, if anyone can produce satisfactory evidence, that this expectation, in a crude form at least, is innate. This is of no logical importance, however; the really important point is that it is not *a priori*, that it is perfectly conceivable that the universe might not answer to these demands and that no such amended definitions that might be suggested would help us.

Now it will be found that the Law of Causation, as actually used, is such that if it be true the world does consist of permanent substances of a few ideal kinds, in a perfectly reasonable sense of *permanence* and *kind* which is only an extension of our original senses of these words. The Law of Causation says that every event has a cause. It refers to definite particular events and to each one ascribes another definite event or set of them as *its* cause. What then is meant by *a cause*? Evidently it has something to do with causal laws, but the precise connexion is not at first obvious. Causal laws, even in their crudest form, connect, not definite particular events, but classes of abstract events. For they imply the possibility of recurrence under varying conditions and at different times and places. Even the crudest sort of causal law is doubly abstract; it takes the form: Whenever an event of the sort η happens to a substance of the sort α an event of the sort η^1 follows after a certain lapse of time t in a substance of the sort α^1. Of course as a particular case η^1 and η may be the same kind of event, α and α^1 may be the same kind of substance, and the two events may happen in the same substance. Again, of course, the antecedent in a causal law may be several abstract events in substances of several kinds; and these events may not be contemporary with each other. The same is true of the consequent. But in any case the important point for us to notice is (a) that the antecedent and the consequent in any causal law are doubly abstract and (b) that the Law of Causation, on the contrary, is an assertion about definite events in definite substances. To use a phrase employed by Mr. Russell in *Principles of Mathematics* the Law of Causation deals with 'the causation of particulars by particulars'; and we have to reconcile this with the fact that no causal law deals with particulars at all.

The way to reconcile the two facts is as follows. We assume that any definite particular event can be unambiguously *described* by mentioning a finite number of abstract characteristics. These together tie us down to one definite substance or set of substances and to one definite event or set of definite events in these substances. Each of the characteristics used in the description is abstract, and, taken by itself, can recur at other times and places and in other substances. Each can therefore be taken (say) as the consequent in some causal law, and the antecedent of each in that causal law will, of course, again be abstract. The further assumption is that these abstract antecedents when taken together will once more suffice to tie us down to a single definite event or to a set of definite events in a single definite substance or set of definite substances. This event or set of events is then *the* cause of the definite event or set of events with which we started.

Thus the Law of Causation, in asserting that every event has a cause, makes the following three assumptions. (i) Every definite event can be unambiguously described by mentioning a finite number of its abstract characteristics. (ii) Either each of these characteristics taken separately, or selections out of them which together exhaust them, are consequents in causal laws. (iii) The antecedents in these causal laws are a set of abstract characteristics which, when taken together, unambiguously describe a definite event or set of events.

19

We have now seen what the Law of Causation asserts; we can now see how it enables us to extend our definitions of *kind* and of *permanent substance*. The individual instances of a kind (even of a first-order kind) do constantly change their first-order properties, and thus at any moment two instances may be in very different states. But all these changes are subject to laws; these are characteristic of the kind, and they do not change. The permanence of first-order properties and their exact similarity among all instances, which first suggested kinds and permanent things, breaks down; but it is replaced by permanence of laws, *i.e.*, of second and higher-order properties. Contemporary states do not now cease to be states of substances of the same kind merely because they differ in their first-order properties; for these differences in first-order properties are compatible with, and indeed are the consequence of, identity of higher-order properties combined with the varying external conditions which are implied by differences of place.

Pari passu with this modification of the notion of a kind goes a change in the notion of the permanence of a given thing. In the first place, even though spatio-temporal continuity throughout a series of states be still demanded as a necessary condition of identity, we no longer demand exact similarity of first-order properties. We are content with permanent laws + reversibility. By this I mean that if S be a certain state of a certain substance we do not demand that every state of a series shall be exactly like S in order to count as belonging to the substance; we admit very different states under different conditions; but we do demand that by suitably reversing the conditions any state that has happened in the series can be reproduced. And we assume that when this condition is not fulfilled we are not dealing with an elementary substance, and that all substances which do not fulfil it are compounded of substances which do fulfil it.

I think that we also demand some kind of first-order identity throughout the series, though it may be very slight, and, to superficial observation, very unimportant and obscure. This is why we make so much of all laws of conservation, *e.g.*, the conversation of mass, of energy, of momentum, and so on.

Corresponding to these changes a new notion is introduced side by side with the old notion of things. This is the notion of the causally isolated system. The old single substances of common sense, determined largely by spatial continuity of matter within a limited region still persist, but the notion of the isolated system composed of several such substances separated in space, largely usurps their place. Such a system is one in which all the laws governing the changes of first-order properties throughout the parts refer only to other parts of the same system and to their spatial relations and not to anything outside the system. An isolated system is thus the old single substance in a much modified and purified form. The importance of continuous filling of a boundary has diminished, and the parts are not series of precisely similar states. But, regarding the system as a whole as a substance spread out in space and time, all its variations follow constant rules and none of these rules refer to anything outside itself. The existence side by side of the new notion of the isolated system and the old criterion of one substance as what fills a certain boundary leads to the distinction between immanent and transeunt causation. The causal laws characteristic of the system are immanent to *it*, as referring to nothing but its parts, but are transeunt to each of its parts, as referring to changes in other parts to account for the changes in any given part.

Complete causal isolation is of course an ideal rather than a fact. What we find is that a system is isolated for certain changes in its parts and for a certain degree of accuracy in accounting for these changes; for other changes and for greater degrees of accuracy different and in general larger systems must be considered. But it is evident that the law of causation would be a useless platitude and that the notions of permanent substance and kind would have broken down beyond hope of salvation if nature were not so constituted that there are systems much smaller than the whole of nature which are for many changes practically isolated.

20

Let me at length sum up the results of this long, confused, and confusing discussion. All particular inductive arguments depend on probability and only lead to probable conclusions, *whatever* we may assume about nature. But *unless* we assume something about nature they give no finite probability to any law (a) because an indefinite number of alternative hypotheses which are not laws are as probable antecedently as the suggested law, and (b) because we are not equally likely to have met with any instance of the class under discussion, since it is quite certain that if there be instances remote in space or time they *could* not have fallen into the selection which we observed. What we actually assume is that nature consists of a comparatively few kinds of permanent substances, that their changes are all subject to laws, and that the variety of nature is due to varying combinations of the few elementary substances. These assumptions are neither self-evident nor mutually independent nor are they capable of complete proof or disproof by experience. The actual course of the process by which we reach these assumptions is somewhat as follows. Nature, even as known to us superficially, presents a surprisingly selective appearance. Of sorts of substances which are *a priori* possible and could be perceived if presented only a very small selection is presented, whilst those sorts which we do meet with have very large numbers of instances. And, to a superficial view even, there are many series of states in nature which have the kind of spatio-temporal continuity which characterises a thing and moreover show practical constancy of first-order properties over long periods of time. Reasons have been given to show that this appearance can hardly be due to limitations of our powers of perception and interest *within* the spatio-temporal field of actual human experience. The view that these characteristics may only be true of a small part of nature into which we happen to have fallen was then discussed. It was argued that, as an objection to the possibility of induction, the argument is unsatisfactory. Either it literally assumes that our connexion with the part of nature with which we are connected is a random one, or that we have arisen here rather than elsewhere because of laws of nature. The latter view *assumes* laws of nature in regions spatio-temporally outside that with which we have come in contact through experience, since the supposed conditions for the origin of human experi*ents* cannot themselves have fallen within the region of nature open to direct human experi*ence*. If, on the other hand, the view that the human race is as likely to fall into one part of the course of nature as into another be taken literally, we can show that it is highly improbable that the general characteristic of confinement to kinds, which we have noticed, extends but slightly beyond the limits of human experience. We thus seem justified in disregarding the possibility that this characteristic of the experienced world does not extend beyond it, as an argument against induction.

Up to this point, however, we can only say that experience has *suggested* a simple ground-plan of the material world to us, and that it is reasonable to suppose that this plan extends beyond what we have actually experienced. So far we have neither formulated the plan in rigid terms, nor, on the face of it, does nature, even as experienced, completely accord with it. At this stage the distinction between elements and compounds and between the perceptible and imperceptible parts of bodies, a distinction itself suggested by much even in the crudest experience, comes to our help. Pursuing this suggestion we have found it possible to regard nature as built up of a comparatively few natural kinds of the first order, all instances of which are *exactly* alike and *completely* permanent. An analysis of the meaning of kinds and of the permanence of substances has shown us what is the precise 'cash-value' of these statements. It has shown that it is because nature, so far as our experience goes, obeys laws in its changes, that the criterion of persistence of substances and sameness of kinds, which broke down when we confined ourselves to first-order properties, can be rendered satisfactory by taking into account second and higher-order properties. It follows that it is a fundamental error to take the scientific notion of substance by itself as 'something that any fellow can understand', and then raise difficulties about the law of causation. The notions of permanent substances, genuine natural kinds, and universal causation are parts of a highly complex and closely interwoven whole and any one of them breaks down hopelessly without the rest.

The upshot of the matter is that whenever we make a particular induction we have this general view about nature at the back of our minds. If we think that we have hold of a substance that is an instance of one of the few fundamental natural kinds, we attach great weight to our induction, otherwise we do not. The logical position is then (a) that those inductions which we regard as highly probable are so relatively to the belief that we really have got hold of the general ground-plan of nature in the region of phenomena under investigation; (b) the evidence for this is never of the nature of a 'knock-down' proof and no numerical probability can be assigned to it. The kind of evidence is that this plan is suggested to us in a rough form by crude experience, and that, as we investigate nature more and more thoroughly, experience itself *suggests* ways in which we can state this plan with greater and greater definiteness and rigour, and, at the same time, nature is found to *accord* with the more rigorous and definite plan far better than it did with the first crude suggestion of a plan. *E.g.*, we believe that we have got very near to the ground-plan of the material world in the theory of chemical elements, in the laws of mechanics, and in Maxwell's equations, and it is relative to these beliefs that particular inductions in chemistry, electricity, etc., are practically certain. The certainty of the most certain inductions is thus relative or hypothetical, and the probability of the hypothesis is not a kind that can be stated numerically.

21

I think that the actual history of the natural sciences bears out this view. They flounder about in the dark till some man of genius sees what are the really fundamental factors and the really fundamental structure of the region of phenomena under investigation. In mechanics the keystone is the notion of acceleration; in chemistry it is the theory of elements and compounds and the conservation of mass; in economics, perhaps, it is the notion of marginal utility. Sciences where no such discovery has yet been made, such, *e.g.*, as psychology and biology are almost at a prescientific level; their inductions carry no great conviction to anyone trained in the more advanced sciences.

At the beginning of the first part of this paper I told the reader that I was extremely doubtful as to the additional principles about nature, which are needed if any law is to be rendered reasonably probable by induction. I have done my best in this second part to indicate the beginnings of an answer to my own question. But I am painfully aware that the article is complex and diffuse without being exhaustive. There is hardly a line in it which I could seriously defend even against myself if I chose to be an hostile critic. But I print it in the knowledge that if I now spend more time I shall only puzzle myself more thoroughly, and in the hope that its very badness may convince the charitable reader at least of the extreme difficulty of the subject.

NOTE

[1] The mathematical theory of the probability of hypotheses is treated by Boole in his *Laws of Thought*. The problem in its most general form (where it is not assumed that h implies $c_1, c_2 \ldots c_n$, but only that it modifies their probability) has been worked out, but not I think published, by Mr. W. E. Johnson. I take this opportunity of expressing the very great obligations which I am under to Mr. Johnson, obligations which I know are felt by all those who have had the privilege of attending his lectures on advanced logic or discussing logical problems with him. Mr. Johnson, however, must not be held responsible for the views expressed in the present paper.

CRITICAL NOTICE:[1]

A Treatise on Probability. By J. M. KEYNES, Fellow of King's College, Cambridge. London: Macmillan & Co. Ltd., 1921. Pp. xi + 466.

Mr. Keynes's long awaited work on Probability is now published, and will at once take its place as the best treatise on the logical foundations of the subject. The present reviewer well remembers going over the proofs of the earlier parts of it in the long vacation of 1914 with Mr. Keynes and Mr. Russell. From these innocent pleasures Mr. Keynes was suddenly hauled away on a friendly sidecar to advise the authorities in London on the *moratorium* and the foreign exchanges. Mr. Russell (like the foreign exchanges) received a shock, from which he has never wholly recovered, in learning that the logic books had been deceiving him by their reiterated assertions that 'man is a rational animal'; and the *Treatise on Probability* was held up till this year.

The present treatise is essentially philosophical rather than mathematical, although it contains a fair amount of mathematics. It is divided into five parts. The first defines probability and discusses how far it can be measured. The second gives the fundamental theorems of probability in strict logical form. This part owes a great deal to Mr. W. E. Johnson, to whose magnificent work on this subject Mr. Keynes acknowledges his great obligations. Indeed the Muse of Probability seems to have fixed her seat at King's College, Cambridge, of which both Mr. Keynes and Mr. Johnson are fellows. The third part deals with the logical principles of inductive and analogical generalisation; and the fifth with the connected, but more complex, problem of inductive correlation or statistical inference. In between these two is sandwiched Part IV, which is entitled 'Some Philosophical Applications of Probability'. This is concerned with a number of historically interesting problems, and in particular with the application of probability to ethics. At the end of the work Mr. Keynes provides an admirable bibliography of books and articles on probability and kindred subjects.

In this review I shall try to give an outline of Mr. Keynes's theory. I shall not have many serious criticisms to make, because I am substantially in agreement with him, and where I am not persuaded by his arguments the subject is so difficult that I have little of value to suggest as an alternative to his views.

The fundamental thesis of the book is that probability is a relation between propositions, which may be compared with implication. When p implies q the belief that p is true justifies an equally strong belief in q. But there are numberless cases where a belief in p justifies a certain degree of belief in q, but does not justify so strong a belief in q as we have in p. In such cases there is a certain logical relation between p and q, and this relation is of the utmost importance for logic. But it is not the relation of implication. It is this other relation with which probability is concerned. This probability relation is capable of degree, since it may justify a more or a less confident belief in q. The typical probability statement is of the form 'p has to q a probability relation of degree x'. Implication may perhaps be regarded as the strongest probability relation, or better as a limit of all possible probability relations.

There is however a very important difference, which is not merely one of degree, between the implicative and the probability relations. There is nothing corresponding to the Principle of Assertion in probability. If one proposition implies another and we know that the first is true we are justified by the Principle of Assertion in going on to believe the second by itself, and in dropping all reference to the first. We can never do this in probability. We can never get beyond statements of the form 'p has such and such a probability with respect to the datum q'. Propositions are true or false in themselves, though we may need to know their relations to other propositions in order to *know* whether they are true or false. But probability is of its very nature relative. When we talk of *the* probability of a proposition this phrase is always elliptical, as when we say that the distance of London is 120 miles. We simply assume that the person to whom we are speaking will supply from his own mind the same data as we are taking. Two important consequences flow from this. In the first place, a proposition may be highly probable with respect to certain data and yet be false. Its turning out to be false makes no difference whatever to the fact that it is highly probable with respect to these data. Secondly, one and the same proposition may have many different probabilities at the same time, so long as the data are different in each case. In particular a proposition may be highly probable with respect to a certain set of data and highly improbable with respect to another set of data which includes the first set as a part. Thus, if the only fact that you know about a man is that he has recently swallowed arsenic, it is highly probable with respect to these data that he will be dead in the next half hour. If you afterwards get the additional piece of information that he has taken an emetic, the probability that he will die in the next half hour, on the combined data, is much smaller. Neither probability is in any way more 'correct' than the other. This essential relativity of probability is absolutely fundamental, and most previous expositions have suffered by failing to grasp it.

To express these facts Mr. Keynes takes over a useful symbol from Mr. Johnson. He writes $q|p=x$ for 'the probability of q with respect to the datum p is of magnitude x'. Two questions at once arise: (1) Can probability always be measured? and (2) Why do we commonly prefer a probability with respect to wider data to a probability with respect to narrower data? These questions are dealt with by Mr. Keynes in two chapters in the first part.

(1) Mr. Keynes argues that there is no reason to suppose that all probabilities fall into a single scale. All indeed lie between certain truth and certain falsehood, but there may be innumerable series leading from the one to the other. It is only probabilities that lie in the same course that can be directly compared. Two different courses may cut each other at one or more points, *i.e.*, there may be certain probabilities which are common to several different series. When this happens there is a possibility of indirectly comparing two probabilities in different series by comparing both with one that is common to the two series. But, even when we confine ourselves to the probabilities of a single series, there is no guarantee that we shall be able to set up a consistent system of numerical measures for them. Not every series of comparable magnitudes is measurable. The mathematicians have naturally exaggerated the amount of numerically measurable probability in the world; and, when they came across probabilities that were not comparable, or, if comparable, not

numerically measurable, they passed by and 'thanked God that they were rid of a rogue'. Probabilities are only measurable in the comparatively rare cases where we have a field of possibilities which can be split up disjunctively into exhaustive, exclusive, and equiprobable alternatives. This does happen in games of chance and in the 'bag' problems in which mathematicians exercise themselves, but not in many other cases.

It must be noticed that this view of Mr. Keynes's is much more radical than the view that all probabilities are theoretically measurable, but that in most cases the practical difficulties are insuperable. Mr. Keynes points out that there is one and only one theory of probability on which the latter view is plausible. This is the Frequency Theory, which he proceeds to discuss.

There is something bluff and Anglo-Saxon about the Frequency Theory, which no doubt accounts for its extreme popularity with the Island Race in general and with Prof. Whitehead in particular. Moreover there is a real but rather complex connexion between probability and frequency by way of Bernoulli's Theorem; and the very narrow limits within which that theorem and its converse can be applied have been overlooked by most people, as Mr. Keynes points out in the later parts of the present work. Thus there are many excuses for accepting the Frequency Theory. Mr. Keynes has little difficulty in showing that, in the simple-minded form in which it appears in Venn's *Logic of Chance*, it is unsatisfactory, and that Venn tacitly assumes in many places a sense of probability other than that which is laid down in his definitions. Prof. Whitehead's form of the theory, as might be expected, is a good deal more subtle. Unfortunately it is not easy to make out exactly what it is. Mr. Keynes states it in the way in which he has understood it from private correspondence, but admits that he may be mistaken about Whitehead's meaning. It is therefore hardly profitable for a third person to discuss this form of the theory. But it is open to a reviewer to point out what seems to him to be a fallacy in Mr. Keynes's arguments against the theory. Keynes argues that Whitehead's form of the theory shares with Venn's the defect that it cannot satisfactorily explain the fundamental axiom connecting the probability of a disjunctive proposition with the probabilities of its separate parts, *i.e.*, the proposition

$$(p \vee q)|h = p|h + q|h - pq|h.$$

On the Frequency Theory, as interpreted by Mr. Keynes, the datum h determines a certain class α of propositions of which p is a member, a certain class β of which q is a member, and certain classes γ and δ of which the propositions pq and $p \vee q$ are respectively members. The probability of p with respect to h is then defined as the ratio of the number of true propositions in the class α to the total number of propositions in this class. Similar definitions apply, *mutatis mutandis*, to the probabilities of q, pq, and $p \vee q$, respectively. He then points out, quite truly, that the question whether the fundamental addition-theorem mentioned above will hold at all depends entirely on what particular classes, α, β, γ and δ, the datum h does determine for the four propositions in question. So far I quite agree, and think that this is a very serious difficulty in the way of the theory in question. [...]

If the measurement and comparison of probabilities be possible only in a few specially favourable cases it is peculiarly important to be sure what those cases are. This leads to the question: When may we judge two probabilities to be equal? And this leads us at once to one of the *cruces* of the Theory of Probability, *viz.*, the famous *Principle of Non-Sufficient Reason*, or, as Mr. Keynes prefers to call it, the *Principle of Indifference*. In the negative and critical part of this chapter Mr. Keynes found most of the work already done for him by Von Kries, one of the few writers on the philosophical side of probability who are really worth reading. Von Kries has already pointed out the absurd results which a lighthearted use of the Principle of Indifference had led to. He did indeed attempt to base on these a positive statement of the proper limits of the Principle; but I am relieved to notice that Mr. Keynes finds the precise upshot of Von Kries's positive theory as hard to grasp as I have always done myself.

By studying the cases where the uncritical use of the Principle of Indifference ends in absurdities Mr. Keynes elicits the following conditions which must be fulfilled if it is to be applicable. (1) The various alternatives under consideration must be capable of being put into the same form, *i.e.*, they must simply be different instances of a single propositional function ϕ. This cuts out the wild applications of the Principle to pairs of contradictory alternatives in which Jevons habitually indulged. The two alternatives 'x is red' and 'x is not red' are not of the same form. The first means that x has the colour red. The second certainly does not mean that x has the colour 'non-red', for non-red is not a colour. (2) The alternatives must not be sub-divisible into other alternatives of the same form as themselves. Given that x is an inhabitant of Europe it follows that he lives either in Great Britain or in France or in Germany or ... These alternatives are of the same form, and so far all is well. But each of them is divisible into sub-alternatives of the same form as itself. The alternative that x lives in Great Britain is divisible into such alternatives as that he lives in England, that he lives in Scotland, etc.... It is by ignoring this condition that mathematicians who treat of geometrical probability so often reach different solutions of the same problem.

Subject to these two conditions Mr. Keynes states the Principle as follows. The alternatives $\phi(a)$ and $\phi(b)$ are equally probable with respect to the data h, provided that h can be written in the form $f(a)f(b)h'$, where $f(a)$ and $f(b)$ are logically independent, h' is absolutely irrelevant to both alternatives, and $f(a)$ and $f(b)$ are the only parts of h that are relevant to $\phi(a)$ and $\phi(b)$ respectively. [...]

It will be seen then that all judgments of indifference involve judgments of irrelevance. We have to know what part of h is irrelevant to both $\phi(a)$ and $\phi(b)$ before we can see whether h does fall into the form required for the Principle of Indifference. These judgments of irrelevance are of fundamental importance in Probability, and no rules can be given for making them. In the end we have to come down to direct insight, just as we have to do in the end in judging the validity of any deductive argument.

Mr. Keynes makes one very important observation here on the dangers of symbolism. So long as we are dealing with mere a's and b's all that we know about them is that they are both instances of some ϕ. But the moment you substitute something definite, like Socrates, for a, and something else definite, like Plato, for b, you can no longer assume that the conditions for the Principle of Indifference still hold. The moment you know, not merely that you are dealing with a ϕ, but also know *which* particular one of the ϕ's you are dealing with, you may have fresh relevant information.

Having treated the conditions under which two or more probabilities may be judged to be equal Mr. Keynes turns to the question: 'Under what conditions can one probability be judged to be greater or less than another?' Such comparisons can only be made directly when either (a) we have the same data, and one of the propositions whose probability is sought is a conjunctive containing the other proposition as a part; or (b) when the proposition whose probability is sought is the same in both cases, but the datum in one is a conjunctive which includes the data of the other as a part. Into the exact refinements that are needed here I will not enter. Mr. Keynes shows that, by combining cases (a) and (b), we can sometimes indirectly compare probabilities which do not fall under either rubric.

(2) The prolegomena to the measurement of probability are now completed, and we can turn to another most important question which has already been mentioned. If there is nothing to choose in point of correctness between the probabilities of a proposition with respect to a wider and to a narrower set of data why do we prefer the former probability to the latter? Why do we attach more weight to the low probability of the patient who is known to have taken both arsenic and an emetic dying in the next half hour than to the much higher and equally correct probability of the same event relative to the narrower data that he has taken arsenic? This extremely puzzling question is attacked by Mr. Keynes in a chapter on the *Weight of Arguments*. I do not know of any other writer who has raised it except myself in the chapter on Causation in *Perception, Physics, and Reality*; though I do not doubt that Mr. Johnson has an elaborate treatment of it up his sleeve. Roughly speaking, any increase in the amount of relevant evidence increases the weight of an argument, though it may leave the probability unchanged or may decrease it. We have already seen an example of the latter; let us now consider the former. Suppose we start with a probability $a|h$. A new piece of evidence k may arise, and k may consist of two parts k_1 and k_2, one of which is favourably and the other unfavourably relevant to $a|h$. In that case it is possible that $a|hk = a|h$. Nevertheless the weight of $a|hk$ is greater than that of $a|h$. Mr. Keynes discusses various cases in which weights can be compared; and he considers the relation between weight and what is called 'probable error' in statistics. In general a big probable error is a sign of scanty observations, and therefore of a low weight for one's result. But this correlation is not absolutely invariable. I wish that Mr. Keynes had discussed why we feel it rational to prefer an argument of greater weight to one of less weight. I think that our preference must be bound up in some way with the notion that to every event there is a finite set of con-

ditions relative to which the event is certain to happen or certain not to happen. So long as the evidence is scanty a high probability with respect to it does not make it reasonable to act as if we knew that the event would happen, because it is reasonable to suppose that we have only got hold of a very small selection of the total conditions and that the missing ones may be such as to be strongly relevant in an unfavourable direction. If the probability remains high relative to a nearly exhaustive set of data we feel that there is less danger that the missing data may act in the opposite direction. In fact, what we assume is that a high probability with respect to a wide set of data is a sign of certainty with respect to the *complete* set of relevant data.

This exhausts the main features of Part I. Part II is largely the formal development of the fundamental axioms of probability. Much of it could be accepted by a person who rejected Mr. Keynes's view as to what probability really is. The most exciting theorems in this part are due to Mr. Johnson, whose valuable conception of 'Coefficients of Dependence' is introduced and explained. It is worth while to mention a very plausible fallacy in probable reasoning which is detected and dealt with mathematically by Mr. Johnson's methods. It seems plausible to hold that if k is favourably relevant to $m|h$ and m is favourably relevant to $x|h$ then k must be favourably relevant to $x|h$. It is shown here that this is not in general true; and the two conditions under which alone it is true are elicited. It is fairly easy to illustrate part at least of this fallacy by an example. The fact that a man is a doctor increases the probability that he will have visited smallpox patients, and the fact that a person has visited smallpox patients increases the probability that he will get smallpox. It by no means follows that the fact that a man is a doctor increases the probability that he will get smallpox. For this fact also increases the probability that he is properly vaccinated and that he will take reasonable precautions. And this of course decreases the probability that he will get smallpox. Thus we see that it is not enough that k shall be favourably relevant to something that is favourably relevant to x. It is also necessary that k shall not be favourably relevant to anything that is unfavourably relevant to x. The second condition is more subtle, and I cannot at the moment think of any simple example that would illustrate it. As an example of the power of the Keynes–Johnson methods the reader is advised to look at Chapter XVII, in which Mr. Keynes solves in a few lines problems over which Boole spent pages of algebra, arriving as often as not at results which are certainly wrong.

To the mathematician I should imagine that the most interesting thing in this part would be Mr. Keynes's beautiful treatment of Laws of Error, and his general solution of the problem: What form must the law of error take in order that the most probable value of a measured variable shall be represented by the arithmetic, the geometric, the harmonic, and other means, of the observed values? I know of no treatment of this subject which approaches Mr. Keynes's for clearness and generality. To most readers of *Mind*, however, the chapters of greatest interest will be the earlier ones on the notions of *Groups* and *Requirement*.

Both these notions were first devised by Mr. Johnson to deal with such problems in deductive reasoning as are raised by Mill's attack on the Syllogism and by the apparent paradox about a false proposition implying all propositions and a true proposition being implied by all propositions. Mr. Keynes first explains the applications of the theory, and then proceeds to give his own extension of it to the case of probable reasoning.

A group, so far as I can understand, consists of a set of propositions which must contain some formal principles of inference, and includes in addition all propositions that follow from the fundamental set by the principles which are contained in that set. A group is said to be real if the set of propositions which determine it are all known to be true, otherwise it is said to be hypothetical. It is of course possible for the same group to be determined by several alternative sets of propositions, though a given set necessarily determines a single group. Mr. Keynes and Mr. Johnson are both persuaded of the extreme importance of the theory of groups in the logic of inference. I agree with them to this extent, that the *facts* that the theory of groups takes into account are of vital importance. But it does seem to me that they can all be stated much more simply in other terms; and I have failed to find anything specially important that follows from the group notation and would not have been discovered without it. Possibly I am only exhibiting my ignorance. The essential point that the group theory is meant to bring out is the distinction between what Johnson calls the Logical and the Epistemic factors in inference. The latter is the question of the order in which we get our knowledge. E.g., p implies q provided that either p is false or q is true. So far it is irrelevant how we came to know that this disjunction holds.

But when we say 'if p then q' we mean something more than this. We mean that it is possible to know that p is false or q is true without having to know that p is false or having to know that q is true. And the only way in which we can know such a thing is by seeing that the disjunction is an instance of some formally true hypothetical such as 'if SaP then $\bar{P}a\bar{S}$'. Again, if we want to infer q from p it is obviously necessary to be able to know that p is false or q true before you know whether q is true or not. All this can be and is expressed by Mr. Keynes in terms of the theory of groups; and my only doubt is whether it becomes any clearer or leads to anything further when so expressed.

A proposition has a probability with respect to a set of data h when neither it nor its contradictory falls into the group determined by h. Does this really enlighten us any more than to know (what is equivalent to it) that neither the proposition nor its contradictory must follow logically from the *premises* mentioned in h by the known *formal principles* of deductive logic? On page 131 Mr. Keynes has a formidable definition in terms of groups of the statement that "the probability of p does not require q within the group determined by h". When this definition is unpacked it seems to me to amount to no more than this: You can make a selection h' out of h such that no part of h outside h' will alter the probability $p|h'$ when added to h'; and some part of h outside h' when added to h' will alter the probability of $q|h'$. If this be the right interpretation, it is far easier to grasp than Mr. Keynes's definition in terms of groups.

Not only am I doubtful of the fruitfulness of the group theory, I am also not satisfied that Mr. Keynes's treatment of hypothetical groups is adequate. All groups must, so far as I can see, include in their fundamental set formal principles of inference as well as premises. I quite understand that the premises may be hypothetical. But can we really allow the generating principles to be hypothetical also? Mr. Keynes does not discuss this point, which seems to me to be a very important one for a person who is going to admit hypothetical groups.

Let us next turn to Mr. Keynes's theory of inductive generalisation, which is contained in Part III. It is peculiarly gratifying to me to find how nearly Mr. Keynes's view of the nature and limits of induction agrees with that put forward quite independently by me in two articles in *Mind*. We both agree that induction cannot hope to arrive at anything more than probable conclusions, and that therefore the logical principles of induction must be the laws of probability. We both agree that, if induction as applied to nature is to lead to results of reasonably high probability, nature must fulfil certain conditions which there is no logical necessity why it should fulfil. Finally, we agree as to the nature of those conditions, in general outline at any rate. In some way the amount of ultimate variety in nature must be limited, if induction is to be practically valuable; the infinite variety of nature, as we perceive it, must rest on combinations of a comparatively few ultimate differences. But of course Mr. Keynes's theory is far more detailed and subtle than anything of which I am capable; and it is, so far as I know, the only account of the logic of this process which a self-respecting logician can read with any satisfaction.

The problem of induction boils down to this: We examine n things. They have the r properties $p_1 \ldots p_r$ in common; this is called their total positive analogy. There is also a set of properties $q_1 \ldots q_s$ such that each is present in some of the things and none is present in all of them; this is called the total negative analogy. Both the positive and the negative analogies in any actual case are pretty certain to be greater than the *known* positive and negative analogies, which form the only basis of our argument. Our object is to prove some proposition of the form that everything which has the properties $p_1 \ldots p_m$ has the properties $p_{r-1} \ldots p_r$. It is obvious that this can only be possible if some part of the known analogy is irrelevant. *E.g.*, all the examined instances agree in the fact that we have examined them, that they are confined to certain limits of space and time, and differ from all unexamined instances in these respects. Whenever this part of the known analogy is relevant to the attempted generalisation, it is clear that the attempt is doomed to fail. Thus an essential factor in all inductive generalisations is judgments of irrelevance. Many of them no doubt depend on past experience, but Mr. Keynes holds that there must be a residuum which is *a priori*. The only importance of the Uniformity of Nature is that it is a general principle of irrelevance, which asserts that *mere* differences of date and position are irrelevant. Mr. Keynes raises the question in a note whether this is affected by the Theory of Relativity; but he does not answer his own question. However this may be, it seems to me that the Uniformity of Nature, thus defined, is a mere pious platitude; since – whether space and time be absolute or relative –

no two objects or events ever do differ merely in date or place. Such differences always involve their being in intimate spatio-temporal relations with different sets of objects or events, and these differences cannot be assumed to be irrelevant.

Our generalisation always refers to much less than the known positive analogy. When we argue that all swans are white our generalisation only concerns whiteness and those few properties by which we define a swan. But all the examined swans were known to have many other common properties beside these, and we do not know that these are all irrelevant. All that we positively know to be irrelevant at this stage is the properties in the known negative analogy. We can reduce the dangers thus involved by seeking other instances which increase the known negative analogy. For this purpose mere *number* is unimportant. One instance which is known to differ from the previously examined ones in many of those properties which the generalisation assumes to be irrelevant is of more importance than dozens of instances which are exactly like those already examined. But there remains a danger due to the fact that the total analogy is almost certain to be greater than the known positive analogy. The extra and unknown analogies may be relevant; and, since we do not know what they are, we do not know where to look for negative analogies which will prove them to be irrelevant. In this case the only course is to increase the *number* of instances, trusting that, even though they do not differ in any known respects from those that have already been examined, they will probably between them differ in many of the unknown points of positive analogy from the examined instances. All this however tells us how to diminish the objections to an inductive generalisation. It does not tell us that any inductive generalisation will possess a reasonable degree of probability, even when we have carried out these processes to the utmost. Something more is clearly needed if inductive generalisation is to be trustworthy.

The extra factor is dealt with in the chapter on Pure Induction. It is easy to prove that an hypothesis becomes more and more probable the more mutually independent consequences of it are verified. It is also easy to prove that, if it starts with a finite probability, sufficient verification of mutually independent consequences will make its probability approach as near as we please to unity. The problem that remains is: What justifies us in ascribing a finite antecedent probability to any inductive generalisation? To this Mr. Keynes answers that we are only justified if we assume that all the variety of perceptible properties springs from a comparatively small number of generating properties.

To each generating property there corresponds a large group of perceptible qualities, but we must admit the possibility that the class of perceptible qualities corresponding to ϕ_1 and the class corresponding to ϕ_2 may partially overlap. If so the group common to the two will not tie us down to a single generator. Setting this possibility aside for the moment, we see that if a group α of perceptible qualities is found to be accompanied by a group β there is a finite probability that the complete group $\alpha\beta$ corresponds to a single generator, or that the generators of α include among them the generators of β. If this is so α will not be able to occur without β, and there is thus a finite antecedent probability of the generalisation, on which induction can build. If we allow that a group of perceptible qualities may have a plurality of possible generators this argument breaks down; but if we assume that the plurality of possible generators for every set is finite we can still assign a finite antecedent probability to inductive *correlations*, which assert that the next S, or at least a certain proportion of the S's, will be P.

Mr. Keynes seems to me to be right here; and it is true that this is the kind of assumption that does lie at the back of all our scientific reasoning. I have only two remarks to make. (1) Does the theory of generators add anything to the facts? Would it not be enough to assume that perceptible qualities do tend to occur in bundles? This is the whole cash-value of the assumption, and the doctrine of generators seems to be nothing more than a hypothetical explanation of our assumption. (2) Mr. Keynes holds that there is no circle in saying *both* that no inductive generalisation can acquire a finite probability without this assumption, *and* that the results of induction may make this assumption progressively more and more probable.

It is therefore not necessary that the fundamental inductive assumption should be certain. It is enough if it ever had a finite probability; for all subsequent experience has tended to support it. What Mr. Keynes means is, I think, this: If the world is a system with a finite number of generating properties we might expect to find a good deal of regularity and repetition in it. Now, up to the present, we have found more and more regularity and repetition the more carefully we have looked for them. Thus the actual course of experience has been such as to increase the probability of the inductive hypothesis, provided that it started with any finite probability. This works out in practice to the result that a large part of the confidence that we now feel in any inductive generalisation is due, not to the special evidence for it, but to the enormous and steadily increasing amount of regularity that we have found in other regions. There is, I think, no circle in this. Thus the one fundamental assumption of induction is that we can know somehow that the inductive hypothesis that nature is fundamentally finite has a finite antecedent probability. Mr. Keynes admits that it is very difficult to see how we can know this. It is certainly not an *a priori* principle, self evident for all possible worlds, that every system must depend on a finite number of generators. We can only suppose that in some way we can see directly that this has a finite probability for the actual world. But the epistemology of this is at present wrapped in mystery.

In Part IV many interesting problems are discussed; but I must only glance at them. Mr. Keynes ranges from Psychical Research to *Principia Ethica*, and from the *Argument from Design* to the Petersburg Problem; and he has something illuminating to say about all of them. From the point of view of pure probability the most important thing in this part is the definitions of an objectively chance event and of a random selection. The former is very important in connexion with statistical mechanics, the latter in connexion with most statistical reasoning. A chance event is not one which is supposed to be undetermined. Nor is it always one whose antecedent probability is very small. To throw a head with a penny is a chance event, but its probability is $\frac{1}{2}$. An event may be said to be a matter of chance when no increase in our knowledge of the laws of nature, and no practicable increase in our knowledge of the facts that are connected with it, will appreciably alter its probability as compared with that of its alternatives.

Part V deals with the principles of statistical inference. It is too technical for me to give any complete account of it, so I will confine myself to a very short summary of the most important points in it.

(1) Mr. Keynes considers the conditions under which Bernoulli's theorem holds, and shows that they are so restricted that we can seldom in practice count on their being fulfilled.

(2) He severely criticises Laplace, and particularly his famous Rule of Succession. This occurs in connection with the attempted inversion of Bernoulli's theorem. I agree with Mr. Keynes about this rule, but it seems to me that he is a little unfair to it in one respect. He assumes that it always deals with cases where what is drawn is replaced before the next drawing. On that supposition it is true, as he points out, that the formula only holds as the number of drawings tends to infinity. But the same formulae hold without this restriction when the objects drawn are not replaced. And surely, if the Rule claims to have the slightest application to our investigations of nature, the latter is the right alternative. For we cannot observe the same event twice over, any more than we can draw a counter twice out of a bag if we do not replace it.

(3) On all these subjects Mr. Keynes prefers Bortkiewicz, Tschuproff, Tchebycheff, and Lexis to the classical French school. I am afraid that, with the exception of Lexis, these names are mere sternutations to most English readers; but I suppose we may look forward to a time when no logician will sleep soundly without a Bortkiewicz by his bedside. [...]

(4) About past statisticians Mr. Keynes makes a remark which exactly hits the nail. They never have clearly distinguished between the problem of *stating* the correlations which occur in the observed data, and the problem of *inferring* from these the correlations of unobserved instances. There is nothing inductive about the former; but, as it involves considerable difficulties, the statistician has been liable to suppose that, when he has solved these, all is over except the shouting. Thus the inductive theory of statistical inference practically does not exist, save for beginnings in the works of Lexis and Bortkiewicz. These beginnings Mr. Keynes describes and tries to extend.

There are several misprints in the book beside those that are mentioned in the list of errata. On page 170 the various kinds of h's have got mixed up in the course of the argument. On page 183 it is said that "we require $a|ah_2h_2$," when we really want $a|ah_1h_2$. On page 207 substitute $\phi(z)$ for $\phi(x)$ on the left-hand side of the equation. In the formula at the bottom of page 386 read f' for f in the second factor of both numerator and denominator. [...]

I can only conclude by congratulating Mr. Keynes on finding time, amidst so many public duties, to complete this book, and the philosophical public on getting the best work on Probability that they are likely to see in this generation.

NOTE

[1] *Editor's note:* Only such passages are omitted here as have been made redundant by changes in the later printings of Keynes's *Treatise* or have been left out at Prof. Broad's request.

CRITICAL NOTICE:

Logic, Part II (Demonstrative Inference). By W. E. JOHNSON. Cambridge University Press, 1922. Pp. xx+258. [*Abridged.*]

The second volume of Mr. Johnson's great work on *Logic* deals with demonstrative inference, deductive and inductive. It is perhaps even more interesting than the first volume, on account of the extreme practical importance of its main subject, and also on account of the digressions on such matters as Magnitude and Symbolism. It covers the whole range of mathematical reasoning, and it also deals with those types of argument which Mill tried, not too successfully, to classify in his Inductive Methods. Incidentally it contains almost the only good criticism that has yet appeared on a number of fundamental, but rather technical, points in Russell's *Principles of Mathematics*. [...]

The last two chapters are devoted to what Mr. Johnson calls *Demonstrative Induction*. His treatment falls into two parts; (1) certain types of hypothetical syllogism in which an instantial premise leads to an universal conclusion, and (2) his substitute for Mill's Methods. The typical example of hypothetical argument which Mr. Johnson gives is of the form: "If some S is P then all T is U; but this S is P; therefore all T is U". It is thus an argument whose major is a hypothetical proposition with a particular antecedent and an universal consequent. The other premise is the assertion of a certain instance in accordance with the antecedent. The conclusion is of course the assertion of the universal consequent. Now no one would deny the validity of such arguments; the only question is whether they can be called inductive, even in the wide sense in which induction is defined by Mr. Johnson. In their most general form they hardly can be called inductive, for the conclusion is not a generalisation of the instantial minor. Mr. Johnson next quotes examples in which he alleges that the conclusion really is a generalisation of the instantial minor. One example is: "If some boy in the school sends up a good answer, then all the boys will have been well taught; the boy Smith has sent up a good answer; therefore all the boys have been well taught". I cannot myself see that the conclusion of this is a generalisation of the instantial minor. I should have thought that it was obvious that "All the boys have been well taught" could only be a generalisation of such an instantial proposition as "The boy Smith has been *well taught*", whereas the actual minor is "The boy Smith has *sent up a good answer*". I therefore see no ground for counting even this argument as inductive. In fact the only argument of this type which would be genuinely inductive, in Mr. Johnson's sense, would be of the form: "If some boys in the house have measles, all will have measles; the boy Smith has measles; therefore all the boys in the house will have measles". This is demonstrative and inductive, and not altogether remote from the real facts of life, as housemasters know to their cost.

Mr. Johnson points out that arguments of this kind really are common in science. From what we know of the atomic theory we can say with great probability that "If one sample of Argon has a certain atomic weight, then all samples of Argon will have the same atomic weight". We then find that the atomic weight of a certain particular specimen is 40. And we are justified in concluding that all specimens of Argon will have atomic weight 40, provided our major is correct.

I will end with an account of Mr. Johnson's substitute for Mill's Methods. He sees clearly that Mill was confused as to the nature of the methods. Really they should be purely demonstrative, leading to conclusions which are as certain as their premises. And their premises have to be borrowed from the results of problematic induction. Now Mill hardly distinguished the Method of Agreement from Induction by Simple Enumeration, which is a form of problematic induction. Again, he thought that the ultimate majors of these arguments were very wide general principles, like the Law of Causation. Mr. Johnson points out that they need much more definite and concrete majors before they can be rendered genuinely demonstrative. These majors have to be established by problematic induction, and they take the following form in the simplest case. Certain sets of generic characteristics ('determinables', as Mr. Johnson calls them) determine a certain other generic characteristic. Each determinable is susceptible of a number (finite or transfinite) of specific modifications. *E.g.*, 'colour' is a determinable, and a certain definite shade of red is a determinate under it. And of course each determinate is capable of being exhibited in an infinite number of particular instances. With these preliminaries we can state the kind of major premise which will serve for a demonstrative induction. We need – if I understand Mr. Johnson rightly – in the simplest case, to establish a proposition of the following kind as a premise. (1) In all cases where all the determinables ABCD are present the determinable P is present; and no other determinable (say Q) is present in all these cases. (2) In all cases where the determinable P is present all the determinables ABCD will be found; and there will be no other determinable (say E) common to all these cases. When such a premise has been established the demonstrative induction rests on certain axioms about adjectival determination. Let us see how much freedom this premise allows us. If I interpret Mr. Johnson rightly it is quite possible (1) that we should have *abcdp* and *a'b'cdp*, for instance. (2) It is even possible that we should have *abcdp* and *a'bcdp*. But (3), if this be so, we cannot have *a"bcdp"*. In fact we may here conclude A*bcdp*, *i.e.*, that, although the *presence* of A in some form is necessary to the *production* of *p* yet its *variations* are irrelevant to the *variations* of *p*, so long as BCD have the specific values *bcd*. (4) Even if we have A*bcdp*, we must not conclude that variations of A will be irrelevant to variations of *p* when BCD are not confined to the specific values *bcd*. We may perfectly well have *ab'cdp'* in spite of A*bcdp*. (5) Lastly, if we find that *abcdp* and *a'bcdp'*, then we cannot have *a"bcdp* or *a"bcdp'*; we must have *a"bcdp"*. *I.e.*, if *any* variation of A is relevant to variations of P, while BCD have the specific values *bcd*, *all* variations of A will entail variations of P under the same conditions. But (6), even if this be so, we must not conclude that, when the specific values of BCD are no longer confined to *bcd*, we cannot have such a case as *a"b'cdp*.

In all these arguments it is assumed that the determinables under discussion are 'simplex', *i.e.*, that A, for example, is not really a complex of two or more determinables, say $A_1 A_2$. It is also assumed that ABCD are all independently variable. Taking such a major as this, and supplying it with different sorts of minor from our observations, it is clear that we can arrive at four different types of conclusion, according to the nature of the factual minor supplied. (1) If all are simplex, and *abcdp* and *a'bcdp* then A*bcdp*. (2) If all are simplex, and *abcdp* and *a'bcdp'*, then *a"bcdp"*, where *p"* differs from both *p* and from *p'*. (3) If all be simplex, and *abcdp* and *a'bcdp'* then *a"cdp* must be *b"*, where *b"* differs from *b*. (4) If *abcdp* and *a'bcdp'* and *a"bcdp* then A cannot be simplex but must be of the form $A_1 A_2$.

These four types of argument Mr. Johnson calls respectively the figures of *Agreement, Difference, Composition,* and *Resolution*. The reasons for the first two names are obvious. In the third, after a variation in A has produced a variation in P we find that a further variation in A does not produce the expected further variation in P. We therefore conclude that this variation in A has been *compounded* with and neutralised by a variation in some other factor such as B. In the fourth we have the same sort of facts to explain; but we know that there has been no variation in the other factors, whilst we are not sure that all the factors are simplex. We are therefore forced to *resolve* the factor about whose simplicity we were doubtful into two or more factors.

Mr. Johnson illustrates his Figures and then deals with the more complex and actual case of a determined result involving several determinables PQRS, say. The general principles involved are the same, and will be clear to anyone who has understood the argument in the simpler cases.

I think there can be no doubt whatever that Mr. Johnson's Figures are a great improvement on Mill's Methods, both in logical rigour and in approximation to the actual procedure of scientists. There is, however, one criticism which strikes me. Surely the axioms on which Mr. Johnson bases his Figures wholly ignore the possibility of the laws of adjectival determination sometimes taking a *periodic* form. Suppose it happened that P was so connected with ABCD that

$$P = A \sin(BC + D).$$

Then we should have $p = a \sin(bc+d)$ and $p' = a \sin(b'c+d)$ and yet $p = a \sin(b"c+d)$, provided that $b"$ is and b' is not equal to $b + 2\pi n/c$. Nor is this an outrageous supposition, since electromagnetism mainly rests on laws of this kind.

I have perhaps said enough to show that Mr. Johnson's book is one which no one interested in Logic and Scientific Method can afford to neglect. It contains many controversial points, as any thorough treatment of such difficult subjects must do; but I have no hesitation in saying that it is the best book that has appeared, or is likely to appear for a long time, on the absolutely fundamental questions with which it deals.

MR. JOHNSON ON THE LOGICAL FOUNDATIONS OF SCIENCE

Mr. Johnson's great work on Logic goes steadily forward; and the later volumes increase in general interest owing to their more concrete subject-matter, whilst they have all the technical merits of the earlier ones. It will be remembered that, in Part II, Mr. Johnson considered various processes of reasoning which he called 'intuitive', 'summary', and 'demonstrative' induction. None of these is quite what plain men mean by 'induction'; that process Mr. Johnson distinguished by the name 'problematic'. The present volume[1] is primarily concerned with problematic induction, *i.e.*, it deals with the same kind of questions as Mr. Keynes considers in the third Part of his *Treatise on Probability*. Fortunately for the philosophic world Mr. Johnson holds that problematic induction cannot be understood except in terms of certain *a priori* concepts – roughly, those of cause and substance – and that it cannot be justified except on certain postulates which involve those concepts. Consequently a great deal of the book is taken up with extremely valuable discussions about the categories of cause and substance and their relations to each other and to space and time. [...]

The book starts with a very full Introduction which explains in general terms the concepts, and states the results, which are to be more fully discussed in the later chapters. The rest of the book may be roughly divided into four parts. (1) Chapters I to V, inclusive, and the Appendix may be said to deal with the more purely logical aspects of problematic induction. (2) Chapters VI and VII are specially concerned with the notion of the Continuant, which corresponds approximately with the traditional concept of Substance. (3) Chapters IX, X, and XI are concerned with different kinds of causation, and with the spatio-temporal characteristics of causal processes. (4) Chapter VIII deals with the application of causal notions to Mind. These divisions are not absolutely sharp, for Mr. Johnson holds (a) that cause and substance are not so much two categories as "two aspects of a single process of construction" (p. 98), and (b) that the validity of science depends on certain postulates in terms of cause and substance (p. xviii). Consequently the notion of a continuant cannot be fully grasped without reference to causation, and conversely. And again the validity of problematic induction cannot be adequately discussed without reference to substance, cause, and their relations to each other and to space and time. Nevertheless we must begin somewhere, and I propose to treat the subject-matter of the book in the following order: – (A) The Continuant; (B) Causation; (C) The Logic of Problematic Induction; and (D) The Application of the Notions of Cause and Continuant to Minds.

A. *The Continuant* [2]

....

B. *Causation*

So much has necessarily been said about causation in dealing with the Continuant that we can afford to be reasonably brief. The notion of Causation is introduced in the chapter on *Fact and Law* with which the book begins. Mr. Johnson at once rejects two fairly popular opinions. The first is that the assertion '*p* determines *q*' is really a statement about our minds, *viz.*, that whenever we characterise anything as '*p*' our minds are determined to characterise it further as '*q*'. This Mr. Johnson calls a purely 'epistemic' view of causation. He himself holds what he calls a 'constitutive' view, *viz.*, that there is a relation of a peculiar kind between the fact of being characterisable as *p* and the fact of being characterisable as *q*, and that this holds regardless of minds. It seems to me that Mr. Johnson is here hardly using 'epistemic' and 'constitutive' in his usual senses. On the theory which he rejects there is a real relation of causal determination, only it is supposed to be confined to certain states of mind. On the theory which he accepts this same relation holds also between events which are not states of mind. I should hardly have thought that a mere difference of opinion about the *range of application* of a relation whose *existence* is apparently admitted by both parties could be accurately described as a difference between a constitutive and an epistemic view of causation.

The second view which Mr. Johnson rejects is roughly the theory that causal laws are just statements of *de facto* regularities. He distinguishes between such propositions as 'Anything that was p would be q' and 'Everything that is p is q'. The former he calls 'Universals of Law', and says that they express 'nomic necessity'. The latter he calls 'Universals of Fact'. The former imply the latter, but are not equivalent to them. A 'nomically contingent' proposition is of the form 'A thing might be p without being q', and this must be distinguished from the particular factual proposition 'Some p is not q'. Now causal laws are universals of law and are nomically necessary. The need for this distinction is perhaps most clearly brought out in I § 5. In the first place, the belief that if p were to happen q *would* happen is often the reason why p never does happen, and therefore why the factual universal 'No p is non-q' is true. Thus, it is commonly believed that 'if a person *were to* go to a Royal garden-party in bathing-drawers he *would be* turned out'; and it is for this reason (among others) that no-one *does* go in that costume; whence it is true that 'no-one goes so attired and fails to be turned out', which is the corresponding factual universal. Now, if the general *belief in x* is what causes y to *to be true* it can hardly be maintained that x and y are the same proposition. Secondly, we assert such propositions as 'if the molecules of a gas had no extension it would accurately obey Boyle's Law'. And we know that there are no gases of which this is true. Now the corresponding factual universal would be (when stated in negative terms) 'No gas both has unextended molecules and fails to obey Boyle's Law'. This is of course true, since it is implied by the proposition 'No gas has unextended molecules'. But the latter proposition equally implies the factual universal 'No gas both has unextended molecules and fails to *disobey* Boyle's Law'. There is no inconsistency between these two factual universals, and they are both true. But there certainly is an inconsistency between the two propositions 'If a gas had unextended molecules it would obey Boyle's Law' and 'If a gas had unextended molecules it would disobey Boyle's Law'. Hence it seems necessary to distinguish between the universal of law and the corresponding universal of fact.

What Mr. Johnson does not seem to bring out very clearly is the connexion or lack of connexion between nomic necessity and logical necessity, *e.g.*, between the kind of necessity which belongs to the proposition that 'if a billiard-ball were hit it would move' and the kind of necessity which belongs to the proposition that 'if all S were P all non-P would be non-S'. Dr. McTaggart apparently identifies the two, and a discussion of the subject by Mr. Johnson would have been interesting. I can find only two passages which seem to throw light on his view of this question. In I § 4 he says that a nomic proposition 'expresses a relation between the characters p and q indicative of the nature of the world of reality'. In VI § 3 he draws a distinction between causal laws and 'Formal Universals'. Under the latter head he includes the laws of kinematics and the whole of geometry. Formal Universals are laws which apply to space, time, and motion as such, apart from any question of their concrete filling. They do not apply to existents, if by an 'existent' you mean 'whatever is actually or potentially manifested in space and time'. On the other hand, causal laws apply to the concrete filling of space and time; they presuppose formal universals, but the converse does not hold. It is clear from these passages that Mr. Johnson means to restrict the *subject-matter* of nomic propositions to occurrents and continuants. But this still leaves it uncertain whether he supposes the *necessity* which characterises the nomically, the formally, and the logically universal to be the same or different.

To understand the details of Mr. Johnson's treatment of causation it is necessary to notice his distinction between 'cause-factors' and 'a completed cause' and 'effect-factors' and 'a completed effect'. He holds that the occurrence of any characteristic is causally determined by a *finite* number of other characteristics. *E.g.*, suppose that something in fact has the character e. Then it is certain that there is a finite set of characters, say $c_1 \ldots c_n$, possessed by this thing, and such that anything which had this set of characters would also have e. But it may be that a thing which has $c_1 \ldots c_m$ without $c_{m+1} \ldots c_n$ need not have e, and it may be that the former sub-set can occur without the second. A cause-factor or an effect-factor is apparently a single characteristic. A completed cause of a given effect-factor e is a set of characters $c_1 \ldots c_n$ such that anything that had this set would have e also. A completed effect of a given cause-factor c is a set of characters $e_1 \ldots e_n$ such that anything that had this set would have c also. (I § 4 and V § 1.)

With these definitions it is obvious that there can be plurality of completed causes relative to a given effect-factor and plurality of completed effects relative to a given cause-factor, for this merely amounts to saying that we cannot simply convert an A-proposition. (V § 3.) It does not in the least follow from this that causal laws in terms of *completed* causes and *completed* effects cannot be stated in a reciprocal form. By a process of gradual modification at both ends we may get from an irreversible law, such as 'Anything that was $c_1 c_2 c_3$ would be e_1' to a law of the form 'Anything that was $c_1 \ldots c_n$ would be $e_1 \ldots e_m$ and anything that was $e_1 \ldots e_m$ would be $c_1 \ldots c_n$'. (V § 6.) In fact Mr. Johnson holds that we have not got a causal law properly stated until the following conditions are fulfilled: – (1) All the cause-factors are independently definable and variable; (2) All the effect-factors are so too; (3) None of the effect-factors can be inferred from any selection less than the whole of the cause-factors, and none of the cause-factors can be inferred from any selection less than the whole of the effect-factors. When these conditions are fulfilled the causal law is reversible.

It will be noticed that in this discussion on Plurality in V Mr. Johnson does not give any rule for distinguishing a cause-factor from an effect-factor. If the characteristic with which you start is to be called a *cause*-factor then the set of characteristics which together imply it is to be called a completed *effect*; if it is to be called an *effect*-factor then the set of characteristics which together imply it is to be called a completed *cause*; but why it is to be called a cause-factor in some cases and an effect-factor in others Mr. Johnson does not here explain. The discussion of this point is carried a little further in VI, and runs as follows. Some philosophers have made the cause a property of a continuant and the effect an occurrent. An example would be if we said that gravitation caused the fall of the Campanile. On this interpretation, of course, there is complete lack of homogeneity between cause and effect. (VI § 2.) But Mr. Johnson holds, quite rightly, that the causal relation is primarily between occurrents (*ibid.*), though he maintains that these occurrents must be located in certain specified continuants and that the properties of these continuants must be mentioned in any complete statement of causation. (VI § 4.) Since the completed cause and the completed effect *both* involve the three factors of occurrent, the continuant to which it belongs, and some property of this continuant, it is clear that they cannot be distinguished by the different nature of their constituents. Moreover, Mr. Johnson explicitly says that cause and effect are not *epistemically* distinguishable. The cause can be inferred from an adequate knowledge of the effect just as well as the effect can be inferred from an adequate knowledge of the cause. And he admits that it remains a serious question whether there is anything left by which cause and effect can be *ontologically* distinguished. (VI § 2.)

The question has to be dealt with separately for transeunt and for immanent causation. For the former I think that Mr. Johnson's solution is as follows. Transeunt causation always requires two continuants C_1 and C_2 in some specific relation R to each other. When this relation has been established between C_1 and C_2, the state of C_2 (say) which immediately follows differs in some assignable way from what it would have been if C_2 had been left to itself. We then call the establishment of R between C_1 and C_2 a *cause*-factor in this transaction; and we count the divergence of C_2's immediately subsequent state from what it would otherwise have been as the *effect* of this. Of course at the same time C_1's state may be modified by the establishment of R between C_2 and C_1. But this introduces no difficulty; it is simply a case of reciprocal transeunt causality. (*Introd.* § 7.)

Now, so far as I can see, in all cases where there is a difficulty in distinguishing between cause and effect Mr. Johnson has to appeal in the end to transeunt causality. Two cases arise over immanent causality. (1) We may have simultaneous immanent causality. Mr. Johnson illustrates this from the gas-law, $pv = Rt$, where p, v, and t, stand respectively for the pressure, volume, and temperature of a given mass of gas, and R is a constant. Which of these are you to call the *cause* of the rest? "So long as we are concerned only with immanent causality there is absolutely nothing to determine which ... is to be called cause and which effect" (IX § 4.) Mr. Johnson's solution is as follows. The whole process must be analysed into three stages, of which two are transeunt and one is immanent. He begins by distinguishing between the *external* pressure, volume, and temperature – p_e, v_e, and t_e – and the corresponding *internal* variables – p_i, v_i, and t_i. The former are the weight on the piston, the volume of the container, and the temperature of its walls. The latter are the reaction of the gas on the walls, and its volume and temperature. Suppose now that the experimenter arbitrarily modifies p_e and v_e by pressing down the piston. Then the whole causal process must be analysed as follows. (a) An inward transeunt process in which the internal pressure and volume are modified according to the laws $p_i = p_e$ and $v_i = v_e$. (b) An immanent process in which the internal temperature is modified in accordance with the law $t_i = p_i v_i / R$. (c) An outward transeunt process in which the external temperature is modified in accordance with the law $t_e = t_i$. Now in the transeunt processes there is no difficulty in saying which is cause and which is effect. The arbitrary change of external pressure and volume is the cause in the first, and the change of internal temperature is the cause in the second. Mr. Johnson's rule is that in the immanent process those factors must be taken as *causes* which are *effects* in the previous transeunt process, and those must be taken as *effects* which are *causes* in the subsequent transeunt process.

(2) In XI § 5 Mr. Johnson goes further and asserts that it would be impossible to draw a distinction between cause and effect in purely immanent causation, even when it is successive, and not simultaneous as in the case just considered. We all believe that ontologically the earlier parts of the history of a continuant determine the later parts and not conversely. Yet there would have been nothing to suggest this to us if all causation had been immanent; for, with an adequate knowledge of the nature of a continuant we can infer backwards just as certainly as forwards in its history, so long as it is left to itself. As it is, however, 'an immanent process of causality may be broken in upon from without by an influence which modifies the succeeding manifestations.... After the interruption the relation of the succeeding to the preceding is objectively differentiated from that of the preceding to the succeeding.' (XI §5) I must confess that I do not clearly understand this. Take, *e.g.*, a moving billiard-ball which hits a cushion and rebounds. Consider two successive stages x and x' in its course before the impact, and two successive stages y and y' in its course after the impact. Then (a) x' can be inferred by a purely immanent law from x, and conversely. (b) y' can be inferred from y by a purely immanent law, and conversely. (c) Neither x nor x' can be inferred from y or y', nor conversely, by a purely immanent law. But (d) y or y' can be inferred from x or x' together with a knowledge of the impact; and equally x or x' can be inferred from y or y' together with a knowledge of the impact. I really cannot see where the 'objective differentiation' comes in in all this; everything seems to be perfectly symmetrical.

The only other points that I need mention in Mr. Johnson's doctrine of causation are the following. (1) In successive causation we must not suppose the cause and the effect to be momentary events. They are events of finite duration which are adjoined at a common temporal boundary. Thus the typical statement would be: 'The change from A to B causes the change from B to C'. (VI § 6.) (2) In transeunt causation the cause and the effect are always simultaneous. (XI § 5.) *E.g.*, I suppose, that a weight does not have to stand on a table for any finite time before the upward reaction of the table upon it begins. Immanent causation may be either simultaneous or successive. (3) In transeunt causation the cause and the effect always have to stand in a certain peculiar relation which is not temporal. In physical affairs it is a spatial relation, such as contact. What precisely it is in psycho-physiological causation Mr. Johnson does not distinctly tell us.

C. *The logic of problematic induction*

The validity of all science, according to Mr. Johnson, rests on certain 'postulates'. These are propositions which are accepted assertorically and not merely hypothetically, which are nevertheless not self-evident nor capable of inductive proof, and which involve concepts (such as cause and substance) which are 'not given in experience'. These postulates enter even into the singular perceptual judgments which form the materials that science generalises. (*Introd.* § 3.) Whether Mr. Johnson thinks that any further postulates are needed for generalisation, which do not enter into singular perceptual judgments, and, if so, what they are, is not clear to me.

In inductive generalisation we start by observing certain instances which have some determinate value of P and some determinate value of Q. We then want to know whether All P is Q or All Q is P. The question which of these two generalisations we shall seek to make is determined by whether we have already found that Some P is not Q or that Some Q is not P. Let us suppose that we have found the latter and not the former, so that we are seeking to establish that All P is Q. The kind of evidence that we need is the following. In the first place we confine ourselves to a single determinate p_1 under the determinable P, and we try to observe as variable a collection of instances having p_1 as possible. Suppose we find that they all have a certain determinate value q_1 under the determinable Q. We next examine in turn sets of instances having the determinates $p_2, p_3 \ldots p_n$. We will suppose that all the members of each such set are found to have a certain determinate value of Q, and that these values differ for each set. They might be $q_2, q_3 \ldots q_n$. As we have said, within each set we shall try to vary the instances as much as possible. On the other hand, as between any two sets we want as little variation as possible in all other respects except the values of P which distinguish them. Our observational data are now of the form All observed p_1 was q_1 and All observed p_2 was q_2... and All observed p_n was q_n. The corresponding generalisation will be of the form All p_1 will be q_1 and All p_2 will be q_2... and All p_n will be q_n. Now the vitally important point to notice is that the *whole* of the first set of facts is the evidence for *each separate constituent* of the generalisation. Our evidence for believing that all water will boil at 100 °C. at normal pressure is not simply that all observed *water* has boiled at *this* temperature, but that all samples of each chemical compound which have been observed (*e.g.*, alcohol, ether, chloroform, etc.), have been found to have a characteristic boiling-point under a given pressure. I do not think that this point has ever been brought out so clearly as it is by Mr. Johnson's notation. A second feature to notice is that the evidence falls into two parts, *viz.* (a) that constancy in P is accompanied by constancy in Q in spite of variations in other factors, and (b) that variation in P is accompanied by variations in Q in spite of constancy in the other factors.

On the logical relevance of number and variety of instances Mr. Johnson takes very much the same view as Mr. Keynes. We have to remember that P may be a complex group of determinables ABCD. The great danger of inductive generalisation is the following. All the observed instances that were q_1 may have had *abcd* in common. We may omit to notice *d*, or may notice it and treat it as irrelevant. In that case we are likely to put our generalisation in the form All that is *abc* is q_1. And this may be too sweeping; the proper form being All that is *abcd* is q_1. Strictly, we shall only be safe if we put into our subject *all* that is common to the observed instances. But this is not practicable, and we have to distinguish as best we can between relevant and irrelevant common features. The object of choosing as variable instances within a set as possible is to reduce the common features which we do not include in our subject as much as we can. And the only object of multiplying instances within a set is the *hope* that we shall thereby reduce the common features, even where we cannot be positively *sure* that we are doing so. Mr. Johnson points out that the fact that all the observed instances have fallen into a certain restricted region of time or space may be relevant; not because absolute position in space and time is relevant in any causal law, but because all the instances that fall into such a region may agree in their close spatio-temporal proximity to some particular causal agent, and this fact may be highly relevant.

The following remarks may be made here. (1) This shows the practical futility of Mr. Keynes's form of the 'Uniformity of Nature'. (2) In *all* inductions our observations have in fact been confined to a comparatively small region of time and space. Indeed the temporal limits are ridiculously small, since they are determined by the length of human tradition up to date. This should make us very doubtful of inferences about the remote past or future based on inductive generalisations. (3) We always start an enquiry with pretty definite views about what is likely to be relevant to what. These are based on past experience. Where we lack this basis, as *e.g.*, in Psychical Research, we can attach very little weight to our inductions. (4) Relevance is a matter of degree. The more minute the phenomena that we are investigating and the more accurate we want to make our results the less we can afford to treat as irrelevant. (5) There have been certain features, such as the fixed stars and the present constitution of the solar system, which have been present in all human observations and cannot be varied by us. We therefore cannot tell whether they may not be relevant to all our laws. This does not matter very much when we are predicting only a little way ahead; but it makes predictions to very remote periods, when the structure of the solar system and the positions of the fixed stars may be very different, highly precarious. (6) Mr. Johnson does not apparently discuss another source of weakness which Mr. Keynes notices, *viz.*, that we may put more than we ought into the predicate of our generalisation, and conclude, *e.g.*, that All P is XYZ when we ought only to conclude that All P is X.

It remains to consider the detailed theory of 'Eduction' which is put forward in IV and in the Appendix. From such a premise as 'Certain things which are m are p' the first step that one can take is to 'A certain further thing which is m (*e.g.*, the next one that I meet) will also be p'. This is *Eduction*. The next step is *Induction*; but it branches into two forms according to the nature of the original premise. If *all* the observed things which were m were p, we may proceed to the *Pure Generalisation* that all things which are m will be p; but, if only a certain proportion of the observed things which were m were p, we can only proceed to the *Statistical Generalisation* that such and such a proportion of things which are m will be p. (I have invented the last two terms, because something has evidently gone wrong with Mr. Johnson's nomenclature. What he calls 'Appendix on *Eduction*' is not mainly on Eduction, as defined by him in IV § 1, but is rather on the special kind 'inductive inference whose conclusion is class-fractional' (IV § 3), which I have called 'statistical generalisation'.)

In IV he considers eduction from premises all of which are favourable. The typical argument has three premises and two middle terms, one substantival and the other adjectival. It may be put as follows:

s is characterised by $p_1 \ldots p_m$
$p_1 \ldots p_m$ characterise $s_1 \ldots s_n$
$s_1 \ldots s_n$ are characterised by p
Therefore s is characterised by p.

Mr. Johnson argues, rightly I think, that there is no 'pure induction' and no 'pure analogy'. One can only say that some arguments are more of one type and some more of the other type. The most purely analogical argument must introduce at least two substances which are known to agree in a number of characters; and the most purely inductive argument must introduce at least two characters which are known to belong to all the instances examined in the premise. Mr. Johnson also works out in elaborate detail the three kinds of negative evidence that are favourable to an eduction. There is no need to excite ourselves over these, for a little reflexion and manipulation will show the reader that they can all be reduced to the form already given, provided we are allowed to introduce negative characteristics and to substitute for 's is not characterised by p' the equivalent 's is characterised by non-p'.

Suppose now that we have a set of premises of the kind mentioned above. Then the addition of a further substantive s_{n+1} which has all the m predicates strengthens the conclusion if and only if it has some characteristic which all the others lack (or lacks some characteristic which all the others have). If we allow negative characteristics it does not matter which alternative we adopt; Mr. Johnson adopts the latter. (It is evident that we shall have to define what we are going to mean by a 'characteristic', or this condition will become trivial. I think we shall have to exclude disjunctive characteristics, for instance.) Again, the addition of a further characteristic p_{m+1} which belongs to all the n substantives strengthens the conclusion if and only if it be nomically possible for all the rest to characterise a substantive without this one. (Here I am again departing from Mr. Johnson's actual statements. He says that it is necessary that there shall *be* a substantive which has all the other characteristics and lacks this one. I cannot see that this *factual particular* needs to be true, provided that the corresponding *nomic contingency* holds. If I am right Mr. Johnson has here overlooked his own distinction. But, as his rule is more rigid than mine, no practical harm would come from following it.)

A set of characteristics such that no selection from it causally determines the rest may be called an 'Independency'. (I am again modifying, as above.) A set of substantives, such that any one of them has (or lacks) some characteristic which is lacked (or had) by all the rest may be called a 'Variancy'. We may say then that an eductive argument is not in its correct form unless the substantival middle term is a variancy and the adjectival middle term is an independency. It will not be positively fallacious if these conditions be not fulfilled, but it will appear to be stronger than it really is.

About the Appendix all I can do is, with the utmost respect to Mr. Johnson, to parody Mr. Hobbes's remark about the treatises of Milton and Salmasius: 'Very good mathematics; I have rarely seen better. And very bad probability; I have rarely seen worse'. The subject is too technical for discussion here, and so many friendships have been wrecked on Bernoulli's Theorem and its Converse that I will content myself with saying that I am bound to reject Postulate I on page 183, where Mr. Johnson assumes equi-probability of *ratios* instead of equi-probability of *constitutions*. I know that he has done this with his eyes open, and I must leave him to his colleague at King's. I fear that the High Table there will be rent with dissension over this business. Granted Mr. Johnson's postulate his conclusions follow, and the bit of mathematical reasoning which leads to them is extremely beautiful.[3]

D. *Application of the notion of cause to minds*[4]

....

It is needless for me to praise a book which will obviously become a classic. To the professional logician and metaphysician Mr. Johnson's work is of course indispensable. To the psychologist it offers certain passages which he will do well to read and ponder. And the intelligent scientist who wants to see the best statement which has ever been made of the concepts and postulates which he uses daily, and who is willing to give to a pretty stiff book the same attention which he would bestow on a masterpiece in his own subject, may be most strongly recommended to devote his leisure to this work.

NOTES

[1] *Logic*, Part III (The Logical Foundations of Science). By W. E. Johnson, M.A., F.B.A. Cambridge University Press. XXXVI+192 pp.

[2] *Editor's note:* This part of Professor Broad's review article is omitted here.

[3] Mr. Johnson has lately given up this postulate; substituted a much more plausible one for it; and deduced from the new postulate, by an admirable piece of mathematical reasoning, substantially the same results as he reaches in the Appendix. Unfortunately I cannot accept the new postulate on reflection, though it looks harmless enough at first sight.

[4] *Editor's note:* This part of the article is omitted here.

THE PRINCIPLES OF PROBLEMATIC INDUCTION

1. DEFINITIONS OF INDUCTIVE ARGUMENTS

By *Problematic Induction* I mean any process of reasoning which starts from the premise that all, or a certain proportion of, *observed* S's have had the characteristic P, and professes to assign a probability to the conclusion that all, or a certain proportion of, S's will have this characteristic. It is assumed that no intrinsic or necessary connexion can be seen between the characteristics S and P. Where such a connexion can be seen, the fact that all observed S's have been found to be P can hardly be called a *logical premise*; it is at most a *psychological occasion* which stimulates the observer to intuit the intrinsic connexion between S and P. The latter process is called *Intuitive Induction* by Mr. Johnson, and I do not propose to consider it here. It is generally admitted that, when intuitive induction is ruled out, premises of the kind which we are considering can lead only to conclusions in terms of probability. This has been argued independently by Mr. Keynes and myself, and I am going to assume that it is true.

We can now classify Problematic Inductions as follows: (1) We may divide them up according to the nature of their premises. These may be (1.1) of the form, 'All observed S's have been P', or (1.2) of the form, 'A certain proportion of the observed S's have been P'. Then (2) we may divide them up according to the nature of the proposition whose probability they profess to evaluate. This may be (2.1) of the form, 'All S's whatever are P'; or (2.2) 'The next S to be observed will be P'; or (2.3) 'A certain proportion of the total number of S's are P'. The following are the most important types of Problematic Induction: (A) If we combine a premise of the form (1.1) with a conclusion about a proposition of the form (2.1), we have what I will call *Nomic Generalization*, because it professes to assign a probability to a general *law* from an observed *regularity*. (B) Any argument whose conclusion is about a proposition of the form (2.2) we may call an *Eduction*, following Mr. Johnson. There will be two kinds of eduction, according to whether the premise is of the form (1.1) or of the form (1.2). These may be called respectively *Nomic Eductions* and *Statistical Eductions*. Finally (C), if we combine a premise of the form (1.2) with a conclusion about a proposition of the form (2.3), we get what may be called, in a wide sense, a *Statistical Generalization*. (This term would sometimes be confined to the special case in which the proportion in the premise is *the same* as the proportion which is considered in the conclusion.)

2. THE LOGICAL PRINCIPLES OF THE ARGUMENT

I shall begin by considering artificially simplified cases. These will be of two kinds, *viz.*, (a) the drawing of counters from a bag, and (b) the throwing of a counter whose opposite sides are of different colour. I shall try to state clearly both the principles of logic and probability that are presupposed and the assumptions about equiprobability that are made, and to show exactly where each enters into the argument. The only satisfactory way of doing this is to work out the arguments in detail.

2.1. *Principles of probability and formal logic*

The following are the only ones that are needed: (1) If p and q be logically equivalent propositions, then $p|h$ (*i.e.* the probability of p on the assumption that h is true)$=q|h$, whatever h may be. This may be called the *Principle of Equivalence*. (2) If p and q be any two propositions, then $(p.q)|h = (q|h)(p|qh) = (p|h)(q|ph)$.[1] Here $p.q$, $q.h$, and $p.h$ are conjunctive propositions – *i.e.* respectively, 'p-and-q', 'q-and-h', and 'p-and-h'. This may be called the *Conjunctive Principle*. (3) If p and q be two propositions which cannot both be true, then $(p \vee q)|h = p|h + q|h$. Here $p \vee q$ means the disjunctive proposition 'either-p-or-q'. This may be called the *Disjunctive Principle*. (4) If p be any proposition, and $q_1 \ldots q_n$ be any set of mutually exclusive and collectively exhaustive alternant propositions, then $p. \equiv :$ $pq_1. \vee .pq_2. \vee \ldots .pq_n$. This can be called the *Rule of Expansion*.

2.2. *Bag problems*

In these we shall suppose that there is a bag which is known to contain n counters which are qualitatively indistinguishable except in respect of their colours. They are to be drawn out one by one, the colour is to be noted, and the counter is not to be replaced.

2.2.1. *Nomic eduction applied to the bag*

Suppose that m counters have been drawn, and that all have been found to have a certain colour, *e.g.*, red. What is the probability that the next counter to be drawn will be red?

Let us denote the proposition that the sth counter drawn is red by ρ_s. Let us denote our original information about the contents of the bag and the method of drawing by h. For shortness let us denote the conjunctive proposition $\rho_1\rho_2\cdots\rho_m$ by r_m. Then we are asked to evaluate the probability $\rho_{m+1}|r_m h$.

By the *Conjunctive Principle* $(r_m\rho_{m+1})|h = (r_m|h)(\rho_{m+1}|r_m h)$

(1) $\quad\therefore\quad \rho_{m+1}|r_m h = \dfrac{(r_m\rho_{m+1})|h}{r_m|h} = \dfrac{r_{m+1}|h}{r_m|h}$

Now there might originally have been in the bag either 0 or 1 or 2 or ...n red counters. Let us denote these $n+1$ mutually exclusive and collectively exhaustive alternant propositions by $R_0, R_1, ..., R_n$ respectively. It is evident that r_{m+1} is inconsistent with there having been originally less than $m+1$ red counters in the bag. Therefore combinations such as $r_{m+1} R_1$ vanish.

By the *Rule of Expansion* then we shall get

$$r_{m+1} \equiv\; :r_{m+1} R_{m+1} . \vee \ldots . r_{m+1} R_n$$

$\therefore$ by the *Principle of Equivalence*,

$$r_{m+1}|h = \sum_{s=m+1}^{s=n} (r_{m+1} R_s)|h.$$

By the *Conjunctive Principle* this is equal to

$$\sum_{s=m+1}^{s=n} (R_s|h)(r_{m+1}|R_s h).$$

Now a precisely similar argument will obviously lead to the result that

$$r_m|h = \sum_{s=m}^{s=n} (R_s|h)(r_m|R_s h).$$

THE PRINCIPLES OF PROBLEMATIC INDUCTION

If we now substitute these values in equation (1), we get

$$(2) \quad p_{m+1}|r_m h = \frac{\sum_{s=m+1}^{s=n} (R_s|h)(r_{m+1}|R_s h)}{\sum_{s=m}^{s=n} (R_s|h)(r_m|R_s h)}.$$

We must next evaluate terms of the form $r_{m+1}|R_s h$. By definition $r_{m+1}|R_s h = (\rho_1 \rho_2 \ldots \rho_m \rho_{m+1})|R_s h$. By repeated application of the *Conjunctive Principle* it follows that

$$r_{m+1}|R_s h = (\rho_1|R_s h)(\rho_2|R_s h\rho_1)(\rho_3|R_s h\rho_1\rho_2)$$
$$\ldots (\rho_{m+1}|R_s h\rho_1 \ldots \rho_m)$$
$$= (\rho_1|R_s h)(\rho_2|R_s h r_1)(\rho_3|R_s h r_2)$$
$$\ldots (\rho_{m+1}|R_s h r_m).$$

Now if there were originally s reds, and if any one of the n counters was equally likely to be drawn, it is evident that $\rho_1|R_s h = \frac{s}{n}$. At the next drawing there are $n-1$ counters. If there were originally s reds, and if the one which has been drawn and not replaced was red, there are now $s-1$ red counters. If any one of the $n-1$ remaining counters is equally likely to be drawn at the second drawing, it is evident that

$$\rho_2|R_s h r_1 = \frac{s-1}{n-1}.$$

So, on these assumptions

$$r_{m+1}|R_s h = \frac{s}{n} \frac{s-1}{n-1} \frac{s-2}{n-2} \ldots \frac{s-m}{n-m}.$$

We must now explicitly notice that we have been making an assumption at this point about *Equiprobability*. This I will call the *First Premise about Equiprobability*.

It is evident that, on the same assumption,

$$r_m|R_s h = \frac{s}{n} \frac{s-1}{n-1} \ldots \frac{s-m+1}{n-m+1}.$$

89

INDUCTION, PROBABILITY, AND CAUSATION

If we substitute the values just obtained in the numerator and denominator of the right-hand side of equation (2) we get

$$\text{(3)} \quad \rho_{m+1}|r_m h = \frac{1}{n-m} \frac{\sum\limits_{s=m+1}^{s=n} (R_s|h) \, s(s-1)\ldots(s-m)}{\sum\limits_{s=m}^{s=n} (R_s|h) \, s(s-1)\ldots(s-m+1)}.$$

This is the fundamental formula for Nomic Eduction in the case of drawing counters from bags. It is evident that we can get no further unless we can evaluate the terms $R_s|h$. These are the antecedent probabilities of the various alternative possible original constitutions of the contents of the bag. Now some logicians and mathematicians, notably Laplace, have at this point argued as follows. They have assumed that, when nothing is known about the contents of the bag except that it originally contained n counters qualitatively indistinguishable save in respect of their colours, the $n+1$ possible alternatives – viz., that it contains 0 or 1 or …n reds (e.g.) – are equally probable. And they defend this on the authority of the Principle of Indifference. On this assumption the factors $R_s|h$ cancel out on the right-hand side of equation (3), and we get

$$\rho_{m+1}|r_m h = \frac{\sum\limits_{m+1}^{n} s(s-1)\ldots(s-m)}{\sum\limits_{m}^{n} s(s-1)\ldots(s-m+1)} \frac{1}{n-m},$$

and it can easily be shown that this is equal to $\dfrac{m+1}{m+2}$. This is *Laplace's First Rule of Succession*.

Now this application of the Principle of Indifference has been severely and rightly criticised by Mr. Keynes. A simple way of seeing that it *must* be wrong is to put $m=0$. We then reach the conclusion that, before any counter is drawn from a bag, the probability that the first to be drawn will be red is $\frac{1}{2}$. But exactly the same reasoning will show that the antecedent probability that the first to be drawn will be blue is $\frac{1}{2}$. Since no counter can be both blue and red it would follow, by the Disjunctive Principle, that the probability that the first to be drawn is *either* red *or* blue must be $\frac{1}{2}+\frac{1}{2}$, *i.e.*, that it is certain to be one or the other, and this is plainly absurd. Nor is it difficult to see why Laplace's application of the Principle of Indifference is wrong. A set of counters $c_1 \ldots c_n$ which are all red can arise only in one way; but a set of counters $c_1 \ldots c_n$ in which one is red can arise in n ways, since it can arise through c_1 being red or c_2 being red or $\ldots c_n$ being red. Thus the various alternatives $R_0, R_1, \ldots, R_n$ are not all exactly alike in internal complexity, for each is analysable into various numbers of sub-alternatives. It is therefore illegitimate to apply the Principle of Indifference to them. We must therefore reject Laplace's Rule.

This throws us back on the original question. Can we evaluate the probabilities $R_s|h$? If we cannot, the formula (3), though valid, is useless. Let us try the following way. It is known that any counter has some one colour, including for this purpose black and white as colours. Let us suppose that there are v distinguishable colours, including black and white. Then any counter taken at random is equally likely to have any one of these v colours. Let us call this the *Second Premise about Equiprobability*. It must have one and can have only one. Hence the antecedent probability that any counter chosen at random shall have a certain assigned colour – e.g., red – is $\frac{1}{v}$. Now a set of n counters containing exactly s red ones can arise in ^nC_s ways. A typical case might be written

$$\gamma_1 \gamma_2 \cdots \gamma_s \cdot \bar{\gamma}_{s+2} \cdot \bar{\gamma}_{s+1} \cdots \bar{\gamma}_n.$$

Here γ_s is the proposition, 'the counter C_s is red'; and $\bar{\gamma}_{s+1}$ is the proposition 'the counter C_{s+1} is not red'. It is obvious that the probability of any such typical case is

$$\left(\frac{1}{v}\right)^s \left(1 - \frac{1}{v}\right)^{n-s}$$

since the fact that one counter is or is not red is irrelevant to the question whether any other counter is or is not red. It follows that

$$R_s|h = {}^nC_s \left(\frac{1}{v}\right)^s \left(1 - \frac{1}{v}\right)^{n-s}.$$

If we substitute these values in the formula (3) and do a little straightforward algebra, we get

$$p_{m+1}|r_m h = \frac{1}{n-m} \frac{\sum_{m+1}^{n} \frac{(v-1)^{n-s}}{(n-s)!(s-m-1)!}}{\sum_{m}^{n} \frac{(v-1)^{n-s}}{(n-s)!(s-m)!}}$$

It is easy to prove that the sum in the numerator comes to

$$\frac{v^{n-m-1}}{(n-m-1)!},$$

and that the sum in the denominator comes to

$$\frac{v^{n-m}}{(n-m)!}.$$

We thus reach the extremely unsatisfactory conclusion that

$$p_{m+1}|r_m h = \frac{1}{v},$$

i.e., that although all the counters that we draw are found to be red, the probability that the next to be drawn will be red remains exactly what it was when no counters had been drawn.

2.2.2. *Nomic generalization applied to the bag*

It is now easy to pass from Nomic Eduction to Nomic Generalization about the bag. The question now is: Given that m counters have been drawn, and that all have been red, what is the probability that all the counters in the bag are red? This simply means that we have to evaluate $R_n|r_m h$. By the *Conjunctive Principle* we can write

$$R_n|r_m h = \frac{(R_n|h)(r_m|R_n h)}{r_m|h}$$

But $r_m|R_n h = 1$, since the first m to be drawn *must* be red if all in the bag are red. Hence

$$R_n|r_m h = \frac{R_n|h}{r_m|h}.$$

THE PRINCIPLES OF PROBLEMATIC INDUCTION

Making use of the *Principle of Equivalence* and the *Conjunctive* and *Disjunctive Principles*, we get

$$(4) \quad R_n|r_m h = \frac{R_n|h}{\sum_{s=m}^{s=n} (R_s|h)(r_m|R_s h)}.$$

This is the fundamental formula for Nomic Generalization. On the false Laplacean assumption that all terms of the form $R_s|h$ are equal, it is easy to prove that the fraction on the right-hand side of (4) becomes $\frac{m+1}{n+1}$. This is *Laplace's Second Rule of Succession*. On the true assumption about equiprobability – viz., that any individual counter is equally likely to have one of the v distinguishable colours – it is easy to prove that the fraction on the right-hand side of (4) becomes $\left(\frac{1}{v}\right)^{n-m}$. Thus, even on the true assumption, the probability of the law that *all* the counters in the bag are red does increase with every counter which is drawn and found to be red, though the probability that the *next* counter to be drawn will be red does not increase.

2.2.3. *Statistical generalization applied to the bag*

It remains to consider the most general problem, viz., that of *Statistical Generalization*, for the artificial case of the bag of counters. The problem may be stated as follows. We have drawn m counters and have found that μ of them were red and the rest non-red. What is the probability that there were originally x red counters in the bag? We will denote the proposition that m counters have been drawn and that μ of them are red by $r_{\mu,m}$. Then the probability which we have to evaluate is $R_x|r_{\mu,m}h$. Applying the same principles as before, we easily find that

$$R_x|r_{\mu,m}h = \frac{(R_x|h)(r_{\mu,m}|R_x h)}{\sum_{s=\mu}^{s=n-m+\mu}(R_s|h)(r_{\mu,m}|R_s h)}.$$

The limits of the summation in the denominator are determined by the fact that there cannot have been less than μ reds, since μ reds have been drawn, and there cannot have been more than $n-m+\mu$ reds, since $m-\mu$ non-reds have been drawn.

INDUCTION, PROBABILITY, AND CAUSATION

It remains to evaluate $r_{\mu,m}|R_s h$. The μ reds which have been found in the m counters that have been drawn might have been presented in $^mC_\mu$ different orders. We are justified in assuming, on the grounds of the Principle of Indifference, that any order of presentment is as likely as any other, with respect to the available data. This constitutes the *Third Premise about Equiprobability*. It was not needed in the two previous problems. Hence, to find the required probability, we may take a single typical order of presentment, e.g., $r_\mu \bar{\rho}_{\mu+1} \ldots \bar{\rho}_m$, and multiply the probability of this by $^mC_\mu$. Now it is evident that

$$(r_\mu \bar{\rho}_{\mu+1} \ldots \bar{\rho}_m)|R_s h$$
$$= \frac{s}{n} \cdot \frac{s-1}{n-1} \cdots \frac{s-\mu+1}{n-\mu+1} \left(1 - \frac{s-\mu}{n-\mu}\right) \cdots \left(1 - \frac{s-\mu}{n-m+1}\right).$$

So we finally reach the equation

(5)
$$R_x|r_{\mu,m}h = \frac{(R_x|h)\,x(x-1)\ldots(x-\mu+1)(n-x)(n-x-1)\ldots(n-x-m+\mu+1)}{\sum\limits_{s=\mu}^{s=n-m+\mu}(R_s|h)\,s(s-1)\ldots(s-\mu+1)(n-s)(n-s-1)\ldots(n-s-m+\mu+1)}.$$

This is the fundamental formula for Statistical Generalization as applied to the case of the bag. On the false Laplacean assumption that all probabilities of the form $R_s|h$ are equal, it is easy to prove that the most probable value of x is such that $\dfrac{n}{x} = \dfrac{\mu}{m}$, i.e., that the most probable proportion of reds in the whole contents of the bag is the same as the proportion of reds in the set of counters drawn and observed. On the true assumption that

$$R_s|h = {}^nC_s \left(\frac{1}{v}\right)^s \left(1 - \frac{1}{v}\right)^{n-s},$$

it is easy to show that the most probable value of x is such that $\dfrac{x-\mu}{n-m} = \dfrac{1}{v}$, i.e., that the most probable proportion of reds among the remaining

THE PRINCIPLES OF PROBLEMATIC INDUCTION

$n-m$ counters is $\frac{1}{v}$. Now this is exactly what was the most probable proportion of reds in the bag before any counters were drawn, for the antecedently most probable number of reds is that value of s which makes

$$^nC_s \left(\frac{1}{v}\right)^s \left(1 - \frac{1}{v}\right)^{n-s}$$

a maximum, and this is the nearest integer to $\frac{n}{v}$. So, no matter how many counters have been drawn, and no matter what may have been the proportion of reds found among them, the proportion of reds which was *antecedently* most probable for the whole contents of the bag will still be the most probable proportion of reds in the remainder. It would be hard to imagine a less satisfactory result.

2.2.4. *Summary*

To sum up. We have seen exactly how the formal principles of probability and logic enter into the inductive arguments about the bag. We have seen that in *every* case *two* different premises about equiprobability are needed, one in the general course of the argument and another in order to evaluate the terms $R_s|h$. We have seen that, in Statistical Generalizations, a *third* premise about equiprobability is needed. Finally, we have seen that the Laplacean assumption for evaluating $R_s|h$ is certainly false, and that, when the true assumption is made, the inductive argument fails to establish any high probability.

2.3. *Problems on throwing a counter*

We here suppose that there is a single counter which is geometrically regular, and has a red face and a white face. I shall suppose that, for all we know at the outset, this counter may be loaded to any extent either in favour of red or in favour of white.

We must first define the notion of 'loading'. I shall say that 'the counter is loaded to a degree s in favour of red', if, and only if, the antecedent probability of its turning up red would be s for anyone who knew in detail how it was constructed. I will denote this proposition by R_s. It is evident from the geometrical fairness of the counter and our complete

INDUCTION, PROBABILITY, AND CAUSATION

absence of information as to its loading that $R_s|h = W_s|h$. Again, if the counter be so constructed that the antecedent probability of its turning up red is s for anyone who knows its construction, it is evident that the antecedent probability of its turning up white is $1-s$ for the same person. Hence $R_s|h = W_{1-s}|h$. We can therefore confine ourselves to loading for red, for we shall cover all the probabilities if we let s range from 0 to 1. That is,

$$\sum_0^1 R_s|h = 1.$$

2.3.1. *Statistical eduction applied to the counter*

Let us now suppose that this counter is thrown n times, and that, on m occasions, it is found to turn up red. What is the probability that the next throw will be red? It is easy to prove by exactly the same methods as we used for the bag that

$$(6) \quad p_{n+1}|r_{m,n}h = \frac{\sum_0^1 (R_s|h) s^{m+1}(1-s)^{n-m}}{\sum_0^1 (R_s|h) s^m (1-s)^{n-m}}.$$

This is the formula for Statistical Eduction in the case of the counter. More complicated formulae could be got for a geometrically regular die with v sides and a different colour on each, but the principles and premises would be of exactly the same kind.

2.3.2. *Statistical generalization applied to the counter*

The most general formula would be that for Statistical Generalization. Here we suppose that the counter has been thrown n times, and that red has turned up m times. It is now to be thrown a further n' times, and we ask: 'What is the probability that red will turn up m' times in these further n' throws?' It is easy to prove by the same methods as before that

$$(7) \quad r_{m+m', n+n'}|r_{m,n}h = {}^{n'}C_{m'} \frac{\sum_0^1 (R_s|h) s^{m+m'}(1-s)^{n+n'-m-m'}}{\sum_0^1 (R_s|h) s^m (1-s)^{n-m}}.$$

3. THE CAUSAL PRE-SUPPOSITIONS OF THE ARGUMENTS

Now these formulae are precisely analogous to those which would be got for the case of drawing counters from a bag, on the supposition that each counter drawn is *replaced* before the next draw. The notion of loading, however, brings out a fundamental pre-supposition of all inductive arguments, which, though really equally present in the case of experiments with bags, is there more likely to be overlooked. The notion of loading is the notion of a constant cause-factor which operates throughout the whole series of throws and co-operates with other and variable cause-factors to determine the result of each throw. Similarly, if each counter is replaced after it has been drawn from the bag and before the next draw is made, the original constitution of the contents of the bag is a constant cause-factor which operates throughout the whole series of drawings and combines with other cause-factors which vary from draw to draw to determine the actual result of each draw. In the case where the counters are not replaced after each draw we have not indeed a *constant* cause-factor; but we know how the original cause-factor, whatever it may have been, has been altered by the results of the previous drawings.

It might perhaps be suggested that there is one fundamental logical difference between the problems on drawing counters from a bag and the problems on throwing a single counter. It might be said that, in the former, we had to use the *First Premise about Equiprobability,* and that, in the latter, it is not used. I think that this is a mistake due to an inadequate analysis of the notion of loading in the latter problems. In the bag-problems the *First Premise about Equiprobability* is needed in order to pass from the datum that there is such and such a proportion of reds in the bag to the probability that the next counter drawn will be red. Now, of course, there is nothing directly analogous to this in the counter problems. But consider the notion of loading. I defined the statement that 'a counter is loaded to degree s in favour of red' to mean that it is so constructed that, relatively to a knowledge of the details of its construction, the probability of its falling with the red side upward is s. Now how could one pass from the knowledge of its structure to the probability of its falling with the red side upwards? The relevant point about its structure would be the position of its centre of gravity with respect to its geometrical centre. We should then have to consider all the possible angles which the plane of the counter could make with the table at the moment of contact, and to find in what proportion of these a counter with its centre of gravity in the given position would inevitably fall over with the red side upwards. But this would not enable us to evaluate the probability of such a counter falling with the red side upwards unless we knew the antecedent probabilities of its striking the table at each of the possible angles, and these antecedent probabilities could not be evaluated without some assumption about equiprobability. Thus the alleged distinction between the two types of problem must be rejected.

I think that we are now justified in making the following assertion, which, if true, is very important. Every inductive argument, whether it be a nomic generalization, an eduction, or a statistical generalization, equally presupposes the notion of causal determination. It presupposes the following proposition, which I will call the *Fundamental Causal Premise*: 'The result of each experiment is completely determined by a total cause composed of cause-factors of two different types. (i) A factor which is known to be constant throughout all the experiments, or whose variations, if it varies, are known at every stage. (ii) A very large number of variable cause-factors, each of which is as likely to vary in one direction as in another of all the directions in which it can vary.' The course of every kind of inductive argument is the same. It argues (a) backward from the actual results to the present probabilities of the various alternative possible cause-factors of the first kind, and (b) forward from these to the probability of a proposed future result. If this be true it is important for several reasons. (i) There are people who profess to reject the notion of causal determination and yet to believe in the validity of some inductive arguments. Some of them think that this position is consistent, provided they content themselves with Eduction and Statistical Generalization and do not attempt Nomic Generalization. If I am right, this is a complete mistake. You must hold either that there is something in causation, or that there is nothing in induction. To reject causation and accept induction is not, as is commonly supposed, hard-headed; it is merely muddle-headed. (ii) On the other hand, there are people who think that, if we could only be sure of the Law of Universal Causation, all the troubles of induction would be over. This is a profound mistake, for the following reasons: (a) Even if we knew that the *Fundamental Causal Premise* mentioned above is true, we should still be faced with the question whether any inductive argument can establish a respectable probability for any proposition about as yet unobserved things or events. In our arguments about bags and counters we have assumed this premise, but we have reached only miserably low probabilities. It is thus evident that this premise, though necessary, is not sufficient to justify the claims of induction to make some propositions about unobserved things or events highly probable. (b) The Law of Causation is not equivalent to the *Fundamental Causal Premise*. In one respect it is more sweeping. We do not need to assume that *every* event is completely determined by causes. All that we need to assume is that the result of each of our experiments is completely determined by causes. This, however, is not logically important, for knowledge of the general principle would guarantee the particular application; and the general principle might be self-evident, while the particular case, apart from reference to the general principle, might not be self-evident. The really serious objection is that, in another respect, the Law of Causation is not determinate enough. It is not enough to know the general fact that the result of each of our experiments is causally determined. We need to know the more specific fact that it is determined in the particular way mentioned in the *Fundamental Causal Premise*. We need to know that we are in presence of a constant cause-factor, or in presence of a cause-factor whose variation from experiment to experiment is known.

4. CONDITIONS FOR A HIGH FINAL PROBABILITY

Still confining our attention to artificial problems, we can now raise the question: 'What further premises would be needed in order that the argument may give a *high* probability to propositions about unobserved things or events?' We see at once that the trouble always arises over the antecedent probabilities of the various permanent cause-factors, *i.e.*, over terms of the form $R_s|h$. Laplace, by making the preposterous assumption that all these are equal, made them cancel out, and arrived at conclusions which are much too good to be true. We, by making what seems to be the only assumption about equiprobability that is reasonable for the colours of counters of whose origin nothing is known, were able to give certain values to the terms $R_s|h$; but, by so doing, we arrived at probabilities which are almost beneath contempt. In the case of a die or counter of unknown origin and construction, it is difficult to see that there is any reasonable principle on which the antecedent probabilities of the various possible degrees of loading can be assigned. Here the Laplacean assumption is not so obviously absurd as in the case of counters in a bag. For a given degree of loading is not *prima facie* analysable into a group of a certain number of equiprobable sub-alternatives, as a given proportion of red counters in a bag is. It is not unreasonable to say that, if nothing is known of the construction of a die or counter, any kind and degree of loading is as likely as any other; and, on this supposition, the Laplacean Rules of Succession follow easily from our formulae (6) and (7); for we may reasonably assume that the probability of any exact degree of loading s is infinitesimal. We may therefore substitute for $R_s|h$ the expression $\phi(s)\,ds$, where ϕ is an unknown function. Formula (6) then becomes

$$(8) \quad p_{n+1}|r_{m,n}h = \frac{\int_0^1 \phi(s)\,s^{m+1}(1-s)^{n-m}\,ds}{\int_0^1 \phi(s)\,s^{m}(1-s)^{n-m}\,ds}.$$

The Laplacean assumption amounts to supposing that $\phi(s)$ is a constant. It is then easy to prove, by means of Γ- and B-functions, that the expression on the right $= \frac{m+1}{n+2}$. If we put $m=n$ this becomes $\frac{m+1}{m+2}$, which is Laplace's First Rule of Succession. The other rules follow in the same way from formula (7) on the same assumption. If we suppose that m and n in formula (8) tend to ∞, $\frac{m+1}{n+2}$ will tend to the value $\frac{m}{n}$. The proposition that

$$(9) \quad \underset{\substack{m\to\infty \\ n\to\infty}}{\mathrm{Lt}}\; p_{n+1}|r_{m,n}h = \frac{m}{n}$$

may be called the *Inverted Bernoulli Theorem*, which is thus a consequence of the Laplacean Assumption in the case of dies and counters.

Now, if the supposition that all degrees of loading for a counter, or all original proportions of red counters in a bag, are *equally* likely enables the formulae of eduction and of nomic generalization to establish reasonably high probabilities, it presumably follows *a fortiori* that any assumption which favours a high degree of loading or a large original proportion of counters of the same colour will act still more strongly in the same direction. Let us call this the *Assumption of Loading*. It is quite distinct from, and independent of, the *Fundamental Causal Premise*. The latter is the assertion that there is a cause-factor of a certain kind operating throughout the whole series of experiments, and it is necessary if *any* inductive argument is to establish *any* probability at all, high or low. The former is an assumption about the relative antecedent probabilities of the various possible cause-factors of the type required by the *Fundamental Causal Premise*. It is required, not to validate inductive arguments as such, but to validate the claims of *some* of them to produce *high* probabilities.

5. TRANSITION FROM ARTIFICIAL TO NATURAL CASES

We have now completed our analysis of inductive arguments, as applied to artificial cases, and have seen exactly what are the constitutive or ontological conditions which must be fulfilled if such arguments are to be both valid and fruitful. It remains to consider whether there is any reason to believe that these conditions are fulfilled in nature. Let us take the case of investigating swans, finding that all observed swans are white, and arguing to the probability that the next swan, or all further swans, will be white; and let us compare this with the artificial cases which we have so far considered. The analogies are as follows: All swans, past, present and future, may be compared to the total contents of the bag. Drawing a counter, noting its colour, and not replacing it, may be compared to catching a swan, noting its colour, and taking care not to count the same swan again among one's data. So far the analogy is complete. But there are many important differences, and all the more obvious of these are unfavourably relevant to induction as applied to nature, in comparison with induction as applied to the artificial case of the bag. (1) The number of swans past, present and future is unknown; but it is almost certainly very great as compared with the number that have been observed up to any given moment. This would be fatal to the attempt to give a high probability to the proposition that *all* swans are white, even if we accepted the Laplacean assumption, for $\frac{m+1}{n+1}$ would be vanishingly small. It seems to me doubtful whether any assumption of loading which had the faintest plausibility would suffice to give a high probability to a law like 'all swans are white', when the evidence is only that all observed swans have been white. On the other hand, the probability that the next swan to be observed will be white might be reasonably large, in spite of the disparity between m and n, if we could accept some assumption about loading much less radical than Laplace's. With Laplace's assumption it is $\frac{m+1}{m+2}$, which is absurdly high. On the assumption that any swan is antecedently as likely to have any one colour as any other, it is $\frac{1}{v}$, no matter how great m may be. In either case it is independent of n. It is thus reasonable to suppose that, with some assumption about loading intermediate between these two, the probability that the next swan will be white would be fairly high if m were fairly large, in spite of the fact that n is incomparably larger than m. (2) The same swan might happen to be observed several times, and to be mistaken for different swans. This will cause us to think that m is larger than it really is. This source of weakness is absent when we are dealing with events, and not with relatively permanent substances, like swans; for the same events cannot be observed twice over by the same observer, and we can generally say with fair confidence whether different people are observing the same or different events. We may perhaps sum up this difficulty by saying that the investigation of substances in nature is intermediate between the case where a counter is never put back after being drawn, and the case where a counter is always put back after being drawn. (3) We come now to the difference which is most serious. In the case of the counters in the bag, we assumed that, at any drawing, any counter then in the bag was equally likely to be drawn, and this was an essential premise of the inductive argument. Now, if the bag be not too large, and does not have pockets in it, and the counters be well mixed, this assumption seems to be justified; but it most certainly is not justified in applying induction to nature. It breaks down for two reasons. (i) Spatially, only a very limited range is open to our observation. There may be swans on other planets, and, if there are, none of them could possibly have been included among our data. (ii) Similar remarks apply to time. Obviously the swans that could be observed up to a given date could not include any swans that began to exist after that date, and it is equally certain that our observations (including the reports of our ancestors) do not include swans that existed more than a few thousand years ago. It is as if the bag were so large that the greater part of its contents could not possibly be reached by us. The result is that the *First Premise about Equiprobability* breaks down, and, as we saw, every kind of inductive argument requires this premise. An attempt has been made to evade this criticism by appealing to the principle that mere difference of spatial and temporal position is irrelevant. Even Mr. Keynes seems to attach some importance to this principle; but, whether it be true or false, it is surely altogether beside the mark; for there is no such thing as *mere* difference of spatio-temporal position. If A is in a different place from B, the things that immediately surround A will differ from those which immediately surround B. If A exists at a different time from B, the things and events which are contemporary with or immediately precedent to A will differ from those which are contemporary with or immediately precedent to B; and no one can assert that a difference in a thing's near neighbours in space and time is always irrelevant to its other properties.

It is clear, then, that there are important differences between any subject of inductive enquiry in nature and the artificial cases for which we have worked out the general theory of inductive argument; and all the differences which we have mentioned are unfavourable to induction as applied to nature. The probabilities which can be reached in the artificial examples are the unattainable upper limits of the probabilities that can be reached by the application of induction to nature. Half, and only half, of this fact has been recognized by most writers on Inductive Logic. They saw the special sources of weakness in the application of induction to nature, and all the various eliminative methods which they have recognized and formulated are simply ways of reducing these sources of weakness to a minimum; but, having exercised themselves in formulating methods of elimination, they thought that they had done all that was required of them. They failed to notice that they had merely reduced certain obvious sources of weakness, and had given no positive theory of inductive reasoning at all.

We can now see clearly that two tasks must be accomplished if the application of inductive arguments to nature is to be valid, and is to lead to reasonably probable conclusions. (i) We must have some reason to believe that something analogous to 'loading' exists in nature, and that certain kinds of 'loading' are antecedently much more probable than others. (ii) We must somehow get over the objection that, since future and remote events could not have been included among our observed data, the *First Premise about Equiprobability*, on which the validity of every kind of inductive argument rests, seems to have broken down. It is evident that three general questions can be raised about inductive inference. These questions may be described as the *logical*, the *ontological* and the *epistemic* question. The *logical* question is to determine the formal character of inductive arguments as such; to state the principles of formal logic and probability which they use, and to see exactly how these enter into the argument; and to discover what premises about equiprobability they require. It includes the further question as to what further premises are required if the argument is to establish, not merely *some* probability, but a reasonably high probability. This problem has now been completely solved, the first part in detail and the second in outline. The *ontological* question is to determine the minimum assumption about the general structure of nature which will guarantee that the conditions, required in order that an inductive argument applied to natural phenomena may establish a high probability, are fulfilled. If this can be solved there will still remain an *epistemic* question. Do we *know* that nature has this general structure? And, if so, how do we know it? Or, if we do not know *it*, do we at least know that it is highly probable? And, if so, how do we know this? It is to these questions that we must now address ourselves. We will begin with the ontological question.

6. THE ONTOLOGICAL QUESTION

The only good treatment of this question with which I am acquainted is contained in Chaps. XXI, XXII and XXXIII of Mr. Keynes's *Treatise on Probability*. Mr. Keynes's theory may be called the *Theory of Generators*. I think that Mr. Keynes's theory is susceptible of improvement in at least two respects. (1) It is stated very briefly, and when one tries to think it out in detail one finds that it is necessary to recognize certain distinctions which Mr. Keynes does not explicitly make, and to deal with certain complications which he does not explicitly consider. (2) I think it is possible to show that generators may be regarded as convenient parameters for stating and working out the theory, but that all that is needed can be accomplished without assuming that they actually exist in nature. I propose, therefore, (1) to begin by assuming the existence of generators, and simply to improve (as I think) the formal exposition of the theory. Then (2) I shall show that the actual existence of generators need not be assumed.

6.1. *Definitions*

Suppose there is a certain set of determinable characteristics $\Gamma_1, \Gamma_2, ..., \Gamma_n$, which are logically and causally independent of each other. This means that any of the $2^n - 1$ combinations, which can be got by taking them one at a time, or two at a time, or ... n at a time, is both logically and causally possible. Let C be another determinable characteristic. Suppose that there is a certain sub-set of characteristics of the first kind, e.g., $(\Gamma_1, \Gamma_2, ..., \Gamma_r)$, and suppose

(i) That anything which had all the characteristics $\Gamma_1 ... \Gamma_r$ would have the characteristic C, and

(ii) That anything which had only a selection from the set $\Gamma_1 ... \Gamma_r$ might lack C.

Then we say that the set $(\Gamma_1 ... \Gamma_r)$ *generates* C. We call $(\Gamma_1 ... \Gamma_r)$ a *generating set* for C, and we call each of the characteristics $\Gamma_1, \Gamma_2, ..., \Gamma_r$ *generating factors* of C.

These definitions obviously leave it possible (1) that a generating set may generate several characteristics, e.g., $(\Gamma_1 ... \Gamma_r)$ might generate C and C'. (2) A characteristic may be generated by several different generating sets – e.g., there is nothing in the definition to exclude the possibility that C is generated by (Γ_{r+1}) and $(\Gamma_2 ... \Gamma_r, \Gamma_{r-2})$ as well as by $(\Gamma_1 ... \Gamma_r)$. All that is excluded is that C should be generated, e.g., by $(\Gamma_1 \Gamma_2)$ as well as by $(\Gamma_1 ... \Gamma_r)$.

I am going to assume, however, until further notice, that the same characteristic C does not in fact have more than one generating set. On that assumption we can talk of *the generator of* C. To say that $(\Gamma_1 ... \Gamma_r)$ is *the* generator of C means that

(i) Anything that had $\Gamma_1 ... \Gamma_r$ would have C, and

(ii) Anything that had C would have $\Gamma_1 ... \Gamma_r$.

It will, of course, still remain possible that some generating sets generate more than one characteristic. The above assumption will be called the *Denial of Plurality of Generators*.

A generating set which contains only one factor – e.g., (Γ_1) – will be called a *set of the first-order*.

A generating set which contains two, and only two, factors – e.g., $(\Gamma_1 \Gamma_2)$ – will be called a *set of the second-order*.

A characteristic which is generated by a generator of the rth order will be called a *characteristic of the rth order*. If we deny plurality of generators, each generated characteristic will be of one and only one order.

The next conception that we need to introduce is that of *fertility*. A generating set is said to be *sterile* if it generates no characteristic. If it generates s characteristics it is said to *have fertility s*. Thus a sterile generating set is one whose fertility is 0.

The fertility of a *generating factor* may be defined as follows: It is the sum of the fertilities of all the generating sets of which it is a factor.

A *generalization* is a universal proposition connecting two mutually exclusive sets of generated characteristics. Thus, the proposition, 'Anything that had $C_1 C_2 C_3$ would have $C_4 C_5$,' is a generalization. A generalization whose subject consists of μ characteristics, and whose predicate consists of ν characteristics, is said to be a 'generalization of the form $g_{\mu\nu}$'.

6.2. Assumptions

Let us suppose that there are N generated determinable characteristics $C_1 C_2 \ldots C_N$, and that they are *logically* independent of each other – i.e., that there is no *a priori* objection to the occurrence of each of the $2^N - 1$ selections that can be made by taking them 1 or 2 or ... N at a time. Let us suppose that there are n generating factors $\Gamma_1 \ldots \Gamma_n$.

We will assume

(i) That each of the characteristics $C_1 \ldots C_N$ is generated by *some* generating set composed of factors selected from the n generating factors.

(ii) That no generating *factor* is superfluous. This means that every one of the n factors is a factor in *some* generating set which generates *some* characteristic in the set $C_1 \ldots C_N$. (It is, of course, quite possible that some generating *sets* may be sterile and generate no characteristic in the set $C_1 \ldots C_N$.)

THE PRINCIPLES OF PROBLEMATIC INDUCTION

(iii) That $N > n$.

(iv) That each of the generated characters is generated by only one generating set. (*Denial of Plurality of Generators*.)

It follows at once that the N generated characteristics fall into $2^n - 1$ mutually exclusive classes (some of which may be null) corresponding to the $2^n - 1$ generating sets. The set of characteristics generated by the rth order generating set $(\Gamma_1 \ldots \Gamma_r)$ may be denoted by $\alpha_{1\ldots r}$, and similarly for the rest.

6.3. *Application to nomic generalizations*

Suppose that a certain thing has been found to have the characteristics $C_1 \ldots C_\mu \, C_{\mu+1} \ldots C_{\mu+\nu}$. What is the antecedent probability of the generalization: 'Anything that had $C_1 \ldots C_\mu$ would have $C_{\mu+1} \ldots C_{\mu+\nu}$'? This is a generalization of the form $g_{\mu\nu}$.

This generalization will be true if, and only if, the factors which are required to generate the predicate set are contained among the factors which are required to generate the subject set. Suppose, *e.g.*, that $C_1 \ldots C_\mu$ require between them for their generation $\Gamma_1 \ldots \Gamma_r$. Then anything that had $C_1 \ldots C_\mu$ would have $\Gamma_1 \ldots \Gamma_r$. Suppose that $C_{\mu+1} \ldots C_{\mu+\nu}$ between them required a selection from $\Gamma_1 \ldots \Gamma_r$. Then anything that had $\Gamma_1 \ldots \Gamma_r$ would have this selection, and anything that had this selection would have $C_{\mu+1} \ldots C_{\mu+\nu}$. Hence the original generalization must be true.

Now the μ subject properties might between them require 1 or 2 or …n generating factors. Let us denote the proposition that they require exactly r generating factors by μ_r. Let us denote the proposition that the ν predicate properties require between them exactly s generating factors by ν_s. It is evident that we need not consider cases in which $s > r$, for it would then be impossible that the generating factors of the predicate should be contained in those of the subject. A typical generalization of the form $g_{\mu\nu}$ would be: 'Everything that had $C_1 \ldots C_\mu$ would have $C_{\mu+1} \ldots C_{\mu+\nu}$'. Let us denote this by $g^{1\,2\ldots\mu}_{\mu+1\ldots\mu+\nu}$. We want to evaluate the probability $g^{1\ldots\mu}_{\mu+1\ldots\mu+\nu}|h$, when h includes the assumptions enumerated in (6.2).

By using the *Rule of Expansion* and the *Principle of Equivalence* we find that

$$g^{1\ldots\mu}_{\mu+1\ldots\mu+\nu}|h = \sum_{r=1}^{r=n} \sum_{s=1}^{s=r} (g^{1\ldots\mu}_{\mu+1\ldots\mu+\nu} \mu_r \, \nu_s)|h,$$

which
$$= \sum_{r=1}^{r=n} \sum_{s=1}^{s=r} [(\mu_r v_s)|h] [g_{\mu+1...\mu+v}^{1...\mu}|\mu_r v_s h],$$

by the *Conjunctive Principle*. Now the probability that the s generating factors required by the predicate are wholly contained among the r generating factors required by the subject is obviously the ratio of the number of ways of choosing s things out of r things to the number of ways of choosing s things out of n things, i.e.,

$$\frac{r!(n-s)!}{n!(r-s)!}.$$

Hence

(10) $$g_{\mu+1...\mu+v}^{1...\mu}|h = \sum_{r=1}^{r=n} \sum_{s=1}^{s=r} [(\mu_r v_s)|h] \frac{r!(n-s)!}{n!(r-s)!}.$$

The antecedent probabilities of *all* generalizations of the form $g_{\mu v}$ will, of course, be equal.

6.3.1. *Effect of the relative values of* n *and* N

So far we have made no use of the assumption that N, the number of generated characteristics, is greater than n, the number of generating factors. I shall now prove that if $n \geq N$, *every* generalization of every form *might* be false, and that if $n < N$ *some* generalization *must* be true. (1) Suppose that $N = n - p$. Call the N generated characteristics $C_1 C_2 ... C_{n-p}$. Then it is evidently possible that (Γ_1) generates C_1 and it only; that (Γ_2) generates C_2 and it only; and that (Γ_{n-p-1}) generates C_{n-p-1} and it only. Then the remaining characteristic C_{n-p} *cannot* require any of the factors $\Gamma_1 ... \Gamma_{n-p-1}$, and it *must* require *all* the factors Γ_{n-p} to Γ_n, for otherwise these factors will be sterile and superfluous. Thus the set $(\Gamma_{n-p} ... \Gamma_n)$ generates only a single characteristic C_{n-p}, and all the sub-sets within $(\Gamma_{n-p} ... \Gamma_n)$ are sterile. It follows that *no* generalization could be true, on this supposition, and this supposition is clearly possible if $N < n$. Therefore, if $N < n$, all generalizations might be false.

(2) Suppose that $N = n$. Then it is plainly possible that all the generated characteristics should be of the first order. If so, each first-order generating set must generate one and only one characteristic, and all other generating sets must be sterile. Under these conditions it is impossible that any

generalization should be true, and these conditions can exist if $N=n$. Therefore, if $N=n$, all generalizations might be false.

(3) Suppose that $N>n$. Two cases will arise, *viz.*, (3.1) that $N>2^n-1$, and (3.2) that $N\leq 2^n-1$. On the first alternative at least one of the generating sets must generate more than one characteristic. In that case *at least two* independent generalizations must be true. For, to take the weakest case that is compatible with the conditions, suppose that only one generating set generates more than one characteristic, and that this set generates only two characteristics C_1 and C_2. Then both the generalizations g_2^1 and g_1^2 must be true.

On the second alternative it is possible that each characteristic is generated by a different generating set, and therefore that each generating set which is fertile generates only one characteristic. In that case there will be no simply convertible generalizations like g_2^1 and g_1^2; but, nevertheless, even in this most unfavourable case, there will be some true generalizations. Suppose, e.g., that $N=n+1$. Either *some* or *none* of the generating sets of these $n+1$ characteristics have a fertility greater than 1. If any do, then there must be some true generalizations. Suppose, then, that each of the $n+1$ characteristics is generated by a different generating set. We then have $n+1$ fertile generating sets, each of unit fertility. The rest of the generating sets are all sterile. It follows that these $n+1$ generating sets must between them take up all the n generating factors, for otherwise some generating factors would be sterile and superfluous. Let us call the generating sets $\gamma_1, \gamma_2, ..., \gamma_{n+1}$. Suppose, if possible, that *every* γ contains a factor not contained in *any* of the remaining γ's. From each γ select such a Γ. The Γ's thus associated with the $n+1$ γ's cannot all be different, for there are only n Γ's in all. So there will be at least one Γ associated in this way with two or more γ's.

But this is self-contradictory. For, if Γ is associated with γ_1 in this way, it will be a member of γ_1 and not of γ_2; whilst if Γ is associated with γ_2 in this way, it will be a member of γ_2 and not of γ_1. Thus the supposition that every γ contains some Γ which is not contained in any of the other γ's leads to a contradiction, and must be rejected. Therefore it is always possible to find *some* selection of γ's from the original $n+1$ γ's, such that it includes all the factors that are included in some other γ. Consequently *some* generalization of the form $g_{\mu\nu}$ *must* be true. It is obvious that this argument applies *a fortiori* when $N=n+p$, where $p>1$.

Now the total number of possible generalizations of the form $g_{N-1,1}$ is N, and if *any* generalizations be true, some of *these* must be true. Thus, the antecedent probability of a generalization of this kind cannot be less than $\frac{1}{N}$. In practice, however, we can be fairly certain that the subject and predicate of our generalization do not together exhaust the total number of generated characteristics, so we have no right to assign so high an antecedent probability to any generalization that we shall actually meet with. I do not see any way of assigning a numerical value to the antecedent probability of a given generalization, even if we know that $N > n$. In order to do so we should have to evaluate the probability $(\mu_r v_s)|h$ in equation (10). This is the antecedent probability that the subject-characteristics $C_1 \ldots C_\mu$ between them require r factors for their generation, and that the predicate factors $C_{\mu+1} \ldots C_{\mu+\nu}$ between them require s factors for their generation. This could not be evaluated unless we made assumptions either about the antecedent probability that a generated characteristic, chosen at random, shall be of such and such an order, or about the antecedent probability that a generating set, chosen at random, should be of such and such fertility. (These probabilities could not, of course, be independent. Any assumption about the one would obviously affect the other.) I do not see any reasonable principle on which such antecedent probabilities could be assigned. It certainly does not seem reasonable to hold that any of the N generated characteristics is *equally* likely to be of the 1st, 2nd, or nth order; and it certainly does not seem reasonable to hold that any one of the $2^n - 1$ generating sets is *equally* likely to have any degree of fertility from 0 to N inclusive. Common sense would suggest that very high-order and very low-order characteristics would be rare, and that very fertile and very infertile generating sets would be rare. The Principle of Indifference would, I think, allow us to suppose that the antecedent probability of a given fertility would be the same for all generating sets *of the same order*; but it would certainly forbid us to assume that it was the same for generating sets of different orders.

This is a most unsatisfactory result. It is of very little interest to know that the antecedent probability of a generalization is *finite*, for this means only that it is greater than 0. What we want to know is that it is not less than *a certain assignable magnitude*, which would presumably be a function of μ, ν, N, n, and the antecedent probabilities mentioned above. Possibly someone with greater technical ability than I may be able to carry the argument to a more satisfactory conclusion, now that the nature of the problem has been made, as I hope it has, quite clear.

It is important to notice exactly what is the force of the condition that $N>n$. This condition is not needed to prove that any proposed generalization has a finite antecedent probability, as can be seen from equation (10), which makes no use of this condition. The condition $N>n$ simply assures us that *some* generalization *must* be true. This, of course, implies that *any* generalization has a finite antecedent probability. But it can obviously be the case that every generalization has a finite antecedent probability, even though it is not *certain* that any of them is true. The theory of generators, without the assumption that $N>n$, assures us that the antecedent probability of any generalization is finite, in the sense that it is greater than 0. The assumption that $N>n$ assures us that it is finite, in the sense that it is greater than a certain number which is itself greater than 0. The trouble is that we cannot evaluate this number without making assumptions about antecedent probability for which there seems to be very little justification.

6.3.2. *Strengthening and weakening conditions for a given generalization*

Let us now consider what circumstances would tend to strengthen the antecedent probability of a generalization of the form $g_{\mu\nu}$. Let us first consider the subject. It is evidently desirable (a) that the subject-characteristics between them should require as many generating factors as possible. For this will increase the probability that the generating factors required by the predicate are contained among those required by the subject. Mere increase in μ, keeping ν fixed, *may* not secure this, for the added characteristics may between them require no generating factors besides those already required to generate the original μ subject characteristics. Still, an increase in μ does increase the probability that the subject requires a large number of generating factors, and does therefore increase the probability of the generalization. If we assume μ and ν to be fixed, then it is evident that the generalization will have the best chance of being true if (i) the subject contains characteristics of a high order, and (ii) the generating sets of the subject-characteristics overlap as little as may be. (b) It is desirable that the generators of the subject-characteristics shall be as fertile as possible, for this will increase the probability that the predicate-characteristics are contained among those which are generated by the generators of the subject-characteristics.

The conditions which the predicate should fulfil are complementary. It is desirable (a) that it shall require as few generating factors as possible, for this will increase the probability that all the generating factors required by the predicate are contained among those which are required by the subject. Mere decrease of v, keeping μ fixed, *may* not secure this, for the characteristics which remain may require for their generation all the factors required by the original v. Still, a decrease in v does increase the probability that the predicate requires only a small number of generating factors, and does therefore increase the probability of the generalization. If we assume μ and v to be fixed, then it is evident that the generalization will have the best chance of being true if (i) the predicate contains no characteristics of a high order, and (ii) the generating sets of the predicate-characteristics overlap as much as possible. (b) The fertility of the generators of the predicate-characteristics does not seem to be relevant.

6.3.2.1. *Tests for the fulfilment of these conditions.* We now see what conditions tend to strengthen the antecedent probability of a generalization. Is there any way of testing whether they are probably fulfilled in a given case? I think there is.

(a) If a characteristic be of high order, it requires the presence of a large number of generating factors in any instance in which it occurs. Suppose, e.g., that C is generated by the set $(\Gamma_1 \ldots \Gamma_r)$. Then in anything that has C there will be all the $2^r - 1$ generating sets that can be formed out of these r factors. Of course, most of these may be sterile, and none of them need be very fertile. Still, the larger r is the more likely it will be that these $2^r - 1$ sets, which are present whenever C is present, generate between them a good many characteristics. All these characteristics will be present whenever C is present. On the other hand, any selection of them which does not include one of those generated by the complete set $(\Gamma_1 \ldots \Gamma_r)$ can be present without C being present. If, then, we find among the subject-characteristics a certain one C, such that, whenever C is present, a certain large group of other characteristics is present, whilst many selections from this group can be present in the absence of C, there is a presumption that C is a characteristic of a fairly high order.

(b) If a characteristic C is generated by a highly prolific generating set, we shall find that there is a certain large group of characteristics, such that, whenever C is present, they are all present, and, whenever any of them is present, C and all the others are present. By making supplementary experiments and observations on these lines, we could presumably determine with fairly high probability whether the conditions required for a given generalization to have a high antecedent probability were fulfilled or not; and we see that the question whether μ is large as compared with ν in this generalization will be of relatively small importance in comparison with the other conditions. The relative magnitudes of μ and ν will merely be the test that we shall have to fall back upon if the other tests are inapplicable or lead to no definite results.

6.4. *Admission of plurality of generators*

Among the assumptions in (6.2) was included the denial of a plurality of generating sets for the same generated characteristic. Let us now consider in outline the result of relaxing this condition. We are to admit now that a generated characteristic C may be generated in some cases by one generating set, e.g., $(\Gamma_1 \Gamma_2)$; in other cases by another set, e.g., $(\Gamma_2 \Gamma_3)$; and in other cases by another set, e.g., $(\Gamma_4 \Gamma_5 \Gamma_6)$. It is evident that there are three possible kinds of plurality of generators to be considered. The various generating sets which generate C may be either (a) all of the same order, e.g., (Γ_1) and (Γ_2); or (b) all of different orders, e.g., (Γ_1) and $(\Gamma_2 \Gamma_3)$; or (c) a mixture, e.g., (Γ_1), (Γ_2), and $(\Gamma_3 \Gamma_4)$. These three kinds of plurality may be described as *Uniordinal*, *Multiordinal*, and *Mixed Plurality*, respectively. It follows from our definition of generation that a pair of generating sets such as (Γ_1) and $(\Gamma_1 \Gamma_2)$ cannot both be generators of a single characteristic C, for to say that $(\Gamma_1 \Gamma_2)$ generates C implies that Γ_1 does not do so and that Γ_2 does not do so. We can, of course, no longer speak of *the* set which generates a given characteristic; nor can we speak of *the* order of a characteristic, unless we happen to know that the only plurality of generators possible for this particular characteristic is uniordinal.

There is another very important distinction to be drawn in connexion with plurality of generators. We start, as before, with an observed object, having a certain N generated characteristic $C_1 \ldots C_N$. We suppose, as before, that each of them is generated by a set selected from a certain n generating factors $\Gamma_1 \ldots \Gamma_n$ which this thing possesses, and that none of these factors is wholly sterile and superfluous. In this particular thing, at this particular time, of course, each characteristic C will be generated by one and only one generating set. But we are now admitting that, in other things, or in this thing at other times, the characteristic C may be generated by other generating sets. Now two cases arise. Are these other sets to be simply *other selections* from the *same n* generating factors $\Gamma_1 \ldots \Gamma_n$? Or are we to admit that other things may have other sets of generating factors, *e.g.*, $\Gamma'_1 \ldots \Gamma'_m$, and that, in them, C may be generated by one or more generating sets selected out of these m different generating factors?

This question is very closely connected with the second difficulty about induction as applied to nature, *viz.*, the fact that it is certain that all the instances that we have observed have fallen within a certain limited region of space and time, whilst we profess to argue to cases which not merely *did* not come, but *could not* have come under our observation. Mr. Keynes's Theory of Generators is not directly addressed to this difficulty, but to the question of something analogous to 'loading' in nature. But, if we want to avoid the present difficulty, we shall have to assume that the only kind of plurality possible in nature is of the first kind and not of the second. We must assume that, in every thing, at every time and place, the characteristics $C_1 \ldots C_N$ are generated by the characteristics $\Gamma_1 \ldots \Gamma_n$, and that the only plurality is a *Selective Plurality*, *i.e.*, consists in the fact that the same characteristic C may be generated in one thing by one set selected from $\Gamma_1 \ldots \Gamma_n$, and in another thing by another set selected from $\Gamma_1 \ldots \Gamma_n$.

6.4.1. *Application to nomic generalizations*

Let us suppose, to simplify the argument, that we need only consider uniordinal plurality of generators. Suppose that one characteristic C_1, in the subject of a generalization, had a plurality of generators, *e.g.*, suppose that (Γ_1) and (Γ_2) are both generators of C_1. Now it might happen that (Γ_1) also generates a certain characteristic $C_{\mu+1}$ in the predicate of the generalization, whilst (Γ_2) does not. If, now, we were to take another thing with the same subject properties, it is quite possible that in it C_1 should be generated by (Γ_2) whilst Γ_1 was absent altogether. Then if (Γ_1) be the only possible generator of $C_{\mu+1}$, the generalization would necessarily break down for this second thing. This shows the effect of admitting plurality of generators as regards the subject of the generalization.

Let us next consider its effect as regards the predicate of the generalization. So far as I can see, it is *never* a *dis*advantage for the predicate properties to have a plurality of alternative generators, whilst it is *sometimes* a positive advantage. Suppose, *e.g.*, that the generalization was 'Everything that had C_1 would have $C_{\mu+1}$'. (a) Provided that C_1 has only one possible generator, everything that has C_1 must have this generator; and, provided that this generator does generate $C_{\mu+1}$, it cannot matter in the least how many more generating sets are also capable of generating $C_{\mu+1}$. (b) Suppose that C_1 is generated both by (Γ_1) and by (Γ_2). If $C_{\mu+1}$ has only one possible generator, *e.g.*, (Γ_1), the generalization will be wrecked, as we saw, by an object in which C_1 is generated by (Γ_2), and in which Γ_1 is not present. But suppose that either (Γ_1) or (Γ_2) will generate $C_{\mu+1}$, then $C_{\mu+1}$ will occur in *any* object that has C_1, and the generalization will be saved.

Thus the correct statement about the effect of a plurality of generating complexes for a single characteristic would seem to be as follows: (a) It is *always* unfavourable to the antecedent probability if the *subject*-characteristics have a plurality of alternative generators; and it is *never* unfavourable if the *predicate* characteristics have a plurality of alternative generators. (b) If any of the subject-characteristics have a plurality of alternative generators it is favourable to the generalization for the predicate-characteristics also to have a plurality of alternative generators.

If a generalization is to have a finite antecedent probability in the only important sense, *i.e.*, if its probability is to exceed a certain assignable number which is itself greater than 0, the following condition would seem to be necessary. There must be a probability greater than a certain number which is itself greater than 0, *either* (a) that none of the subject-characteristics have a plurality of alternative generators, *or* (b) that, if some of them do, the predicate-characteristics have at least as great a plurality of alternative generators. For, in the latter case, there will be a probability, which is finite in the non-trivial sense, that any generator which generates the subject-characteristics is also a generator for the predicate-characteristics.

6.4.2. *Application to eduction*

It is plain that the admission of a plurality of alternative generators for a given generated characteristic weakens nomic generalizations, since additional assumptions are now needed to guarantee that the antecedent probability of the generalization is finite. This is not so if we confine ourselves to eductive conclusions. Suppose the eductive conclusion is that the next thing that we meet which has C_1 will also have C_2. Suppose that C_2 has only one generator, *e.g.*, (Γ_2), whilst C_1 has several alternative generators, *e.g.*, (Γ_1) and (Γ_2). Then, so long as there is a finite antecedent probability that the next thing which has C_1 will have (Γ_2), there is a finite probability that it will have C_2. But, since the number of generators is only $2^n - 1$, the number of alternative possible generators for C_1 cannot exceed $2^n - 1$. Hence this condition is automatically fulfilled without making any fresh assumption. As a matter of fact the condition laid down above is needlessly sweeping. It would not matter how many alternative generators C_1 had, provided that the number of alternative generators of C_2 bore a finite *ratio* to it, *i.e.*, a ratio greater than a certain number which was greater than 0.

It is important to notice that, even if we confined our efforts in induction to establishing *eduction*, and gave up all attempts to establish *generalization* inductively, we should still be *presupposing* the existence of universal laws. For the justification of eduction involves the assumption of generators, and the connexion between a generating set and the characteristics which it generates is a universal law. We should thus be in the odd position that the existence of universal laws is presupposed by all induction, though no inductive argument can assign any finite probability to any law connecting observable characteristics.

6.5. *The elimination of generators*

It is evident that Mr. Keynes thinks of the generated characteristics as qualities like colour, hardness, noise, etc., which we can observe, and that he thinks of the generating factors as their hypothetical physical causes. It seems clear to me that it must be possible to eliminate the hypothetical generating factors, and to state the case wholly in terms of observable characteristics and their relations. I will give a very slight sketch of how this could be done, on the assumption that there is no plurality of alternative generators for a given generated characteristic.

If the theory of generators be true, and the above assumption be made, all the N characteristics which we are concerned with in inductive arguments must fall, as we saw in (6.2), into 2^n-1 mutually exclusive classes, such as $\alpha_1, \alpha_{12}, \ldots \alpha_{12\ldots n}$. Some of these may contain no members. Now, if the theory of generators be true, each of these classes must form what I will call a *Coherent Set*. A *Coherent Set* may be defined as follows: To say that α is a *coherent set* means that it is a set of characteristics such that no member of it can ever occur without all the rest. Any particular coherent set can be defined by means of any characteristic C that falls within it. Thus the coherent set α^c may be defined as follows: It is the set of characteristics consisting of C itself and of every other characteristic which is always present when C is present, and absent when C is absent. Thus $\alpha^c = \hat{X}[X = C . \vee : x \in X \equiv_x x \in C]$.

Now we can evidently drop the notion of generators altogether, and take the notion of mutually exclusive coherent sets of observable characteristics as fundamental. The fundamental assumption will now be that each of the N characteristics falls into some one member of a set of mutually exclusive coherent sets, whose total number is not greater than 2^n-1, where $n < N$. It is obvious that every relation between these sets which could be *deduced* from the hypothesis of generators can be stated, without this hypothesis, as part of the original assumption. This I will illustrate very briefly.

(1) A set β is *subordinate* to a set α if the presence of any characteristic from α is always accompanied by that of some (and therefore of all) of the characteristics in β, whilst the converse does not hold. This is obviously the kind of relation that holds between α_1 and α_{12}, or between α_1 and α_{123}.

(2) A set β is *immediately subordinate* to a set α if β is subordinate to α, and there is no set γ such that γ is subordinate to α and β is subordinate to γ. This is the kind of relation that holds between α_1 and α_{12}.

(3) A set of sets β, γ, δ form an *exhaustive set of subordinates* to α if each is subordinate to α, and whenever a characteristic from each of the sets β, γ, δ is present, one (and therefore all) of the characteristics in α are present. This is the kind of relation that holds between $(\alpha_1, \alpha_{12}, \alpha_3)$ and α_{123}; or between $(\alpha_1, \alpha_2, \alpha_3)$ and α_{123}.

(4) Three sets, α, β, γ, may be so related that none is subordinate to either of the others, but that the presence of a characteristic belonging to any two of them is always accompanied by the presence of one (and therefore of all) the characteristics belonging to the third. This is the kind of relation that holds between $\alpha_{12}, \alpha_{23},$ and α_{31}.

It is clear that all that is necessary in the assumptions which Mr. Keynes makes in terms of generators and their relations to each other, on the one hand, and to the characteristics which they generate, on the other, could be stated in terms of coherent sets and their relations of subordination, etc., to each other. Since the existence of generators implies the existence of coherent sets having these relations to each other, whilst coherent sets might exist and have these relations to each other even if there were no generators, it is obviously advantageous from a purely logical point of view to state the conditions in terms of coherent sets, and to avoid the assumption of generators. From the practical point of view of expounding the theory and drawing remote consequences from it, it is desirable to continue to employ the notion of generators; but they can now be regarded as no more than convenient parameters. They *may* exist, but it is not necessary to suppose that they *do*.

If we allow plurality of alternate generators for a given characteristic the elimination of generators will be a more complicated business; but it is clear that it must be capable of being carried through even in this case.

6.6. *The establishment of functional laws*

There is one other point which it is important to mention before leaving the Ontological Question. I have assumed that the Γ's and the C's are *determinable* characteristics. Mr. Keynes does not explicitly say that they are, but he evidently must intend the Γ's, at any rate, to be determinables and not determinates, for he assumes that the total number of Γ's is finite, and hopes that it may be comparatively small. This assumption would be absurd if the Γ's were supposed to be determinate characteristics, for even a single determinable generating factor might have a large, and even a transfinite, number of different determinate values. But, if the Γ's are supposed to be determinables, so too must the C's be, for it is evident that a set, no matter how complex, of merely determin*able* Γ's would not generate a determin*ate* C. Two consequences follow: (1) It is not nearly so plausible as Mr. Keynes seems to think that the number of C's should largely exceed the number of Γ's. The number of different determin*able* characteristics that we can observe is by no means large; the qualitative variety which we observe in nature is due to the fact that each of the comparatively few observable determinables, such as colour, temperature, etc., has an enormous and perhaps transfinite number of determinates under it. (2) The assumptions that have been made justify only generalizations of the crudest kind, *viz.*, assertions of the form that, whenever certain determinables are present, certain other determinables will be present. Now only the most backward sciences are content with such generalizations. What we want are *Functional Laws*, i.e., laws which will enable us to predict the determinate values of the predicate-characteristics for any given determinate values of the subject-characteristics. To establish such laws, further assumptions have to be made, and something analogous to the *Method of Concomitant Variations* must be used. These assumptions are stated (whether with complete fullness or accuracy I do not here enquire) in Mr. Johnson's treatment of *Demonstrative Induction* in his *Logic*. The only point that I will mention here is that there now arises the possibility of yet another kind of plurality, which Mr. Johnson rules out, quite unjustifiably in my opinion. This is the possibility that *the same* determinate value of the predicate-characteristics may be determined by *several* different determinate values of the subject-characteristics, i.e., that the functional laws of nature may not all be one-valued functions of the variables.

7. THE EPISTEMIC QUESTION

We have now seen what conditions must be fulfilled in nature if inductive arguments are ever to be able to establish reasonably high probabilities. What evidence, if any, have we for supposing that these conditions are in fact fulfilled? Let us call the conditions laid down in (6.2) *The Principle of Limited Variety*, and let us denote it by l. What we have shown is that, if g be any generalization, $g|l \not< \varepsilon$, where ε is a certain number which is greater than 0, but which we have not been able to evaluate.

Now I do not think that anyone would maintain that the *Principle of Limited Variety* has the slightest trace of self-evidence, or that it can be deduced from anything else which is self-evidence. Hence it must be admitted that we do not *know* that l is true. So the next question is: Has l a finite probability with respect to anything that we do know to be true?

Suppose there were certain known facts, f, relative to which l had a finite probability. Suppose further that, if l were true, certain empirical consequences, e, would follow, and that e is found to be true.

INDUCTION, PROBABILITY, AND CAUSATION

Now

$$l|fe = \frac{(l|f)(e|lf)}{e|f}$$

by the *Conjunctive Principle*. But, by hypothesis $e|lf=1$. Hence $l|fe = \frac{(l|f)}{(e|f)}$. Since $e|f$ cannot be greater than 1, $l|fe \not< l|f$; and, if $e|f<1$, $l|fe>l|f$, which is itself supposed to be greater than 0. So, if these conditions were fulfilled, $l|fe$ would be greater than a certain magnitude which is itself greater than 0. The next question then is: Can we find a set of facts, f, and a set of facts, e, such that $l|f>0$, $e|lf=1$, and $e|f<1$?

It seems to me that there is at least one fact which gives l a faint probability by analogy. We do know that we can actually construct out of simple parts of the same nature complicated structures which behave in very different ways, *e.g.*, watches, motor-cars, gramophones, etc. The differences in observable behaviour are here known to be due simply to differences in arrangement of materials having the same properties; and these materials, and the structures formed of them, are parts of the material world. Relatively to this fact it does seem to me that there is a finite probability that the variety of *material* nature at any rate, should arise in the same way. Hence, if f denotes this fact about artificial machines, I should say that $l|f>0$.

Next, it is certain that there is a great deal of recurrence and repetition in nature; and that, up to the present, the more we have looked for it the more we have found it, even when at first sight there seemed little trace of it. I have dealt with this point in detail in my second article on *Induction and Probability* in *Mind*, **29** (1920) 11–45 [present volume, pp. 17–52]. Now, if the Principle of Limited Variety were true, there would be recurrence and repetition in nature; whilst if it were not, there is very little reason to expect that there would be. Hence, if e be this empirical fact, it seems evident that $e|lf=1$ and $e|f<1$. Consequently, there are facts f and e which fulfil the required conditions, and therefore there are facts f and e such that $l|fe$ is greater than a certain number which is greater than 0.

Finally, we have to apply this result to the question of the antecedent probability of any proposed generalization g.

By the *Rule of Expansion* $g. \equiv\, :gl. \vee .g\bar{l}$

THE PRINCIPLES OF PROBLEMATIC INDUCTION

By the *Principle of Equivalence* $g|fe = (gl)|fe + (g\bar{l})|fe$
By the *Conjunctive Principle* $= (l|fe)(g|lfe) + (\bar{l}|fe)(g|\bar{l}fe)$

Whence $g|fe > (l|fe)(g|lfe)$.

Now $g|lfe$ is certainly not less than $g|l$, for the addition of the two facts f and e certainly does not reduce the probability that g shall be true, given that the *Principle of Limited Variety* is true. Hence $g|fe > (l|fe)(g|l)$.

But $g|l$ is greater than a certain number ε, which is itself greater than 0; and $l|fe$ is greater than a certain number η (*viz.*, $l|f$) which is itself greater than 0. Hence $g|fe > \eta\varepsilon$, which is greater than 0.

We see then that any generalization about the material world has a finite initial probability, relative to the known facts that we can construct a variety of differently acting machines from similar materials and that there is a great deal of repetition and regularity in the material world; and this initial probability will increase as we find more regularity and repetition.

Thus a more or less satisfactory answer can be made to the *Epistemic Question*, so long as we confine ourselves to inductive arguments about the material world. But, so far as I can see, we have no ground whatever to trust inductive generalizations about *mental* phenomena; for here there are no known facts analogous to *f*, the fact that we can construct machines of the same materials to act in different ways.

8. SUMMARY OF CONCLUSIONS

Every inductive argument presupposes, beside the general principles of formal logic and of probability, certain assumptions about equiprobability, and what I have called in (3) the *Fundamental Causal Premise*. If it is to establish a high probability, it requires in addition the assumption that 'loading' in favour of a certain one alternative is antecedently highly probable. In the case of induction applied to things and events in nature, these conditions will not be fulfilled unless nature has a certain particular kind of structure, which may be expressed by saying that it answers to the *Principle of Limited Variety*. We stated this principle in terms of the notion of generating factors, and deduced its consequences, first on the assumption that plurality of generators is excluded, and then on the assumption that it is admitted. We also stated the conditions which tend

INDUCTION, PROBABILITY, AND CAUSATION

to strengthen or weaken the antecedent probability of a generalization on the assumption that nature is subject to the *Principle of Limited Variety*, and we gave certain tests for judging whether a given generalization does or does not fulfil these conditions. Then we showed that the notion of generators, though highly convenient, is not essential to the statement of the *Principle of Limited Variety*. The actual existence of generators may be left an open question, and the fundamental notion may be taken to be that of coherent sets of characteristics related to each other in certain ways. We pointed out that, even on the assumption of the *Principle of Limited Variety*, only crude generalizations connecting determinables can be established by induction. To establish functional laws further assumptions about nature are needed. Finally, we said that the *Principle of Limited Variety* is neither intuitively nor demonstratively certain. But there are two known facts about the material world which are so related to it that the antecedent probability of any proposed generalization about material phenomena with respect to these two facts is greater than a certain number which is greater than 0. Lastly we saw that the same argument does not apply to inductive generalizations about mental phenomena. So that, with our present knowledge, we have no good reason to attach any great weight to the conclusions of inductive argument on these subjects.

ADDENDUM TO 'THE PRINCIPLES OF PROBLEMATIC INDUCTION'

In the course of the above paper I did not work out the case of drawing counters one by one from a bag and *replacing* each after it has been drawn and its colour noted, on the assumption that each counter in the bag is antecedently equally likely to have had any one of a certain v alternative colours. This I now proceed to do.

1. *Nomic eduction*

$$r_{m+1} \equiv \, : r_m . \rho_{m+1}$$

$$\therefore \quad r_{m+1}|h = (r_m . \rho_{m+1})|h = (r_m|h)(\rho_{m+1}|h.r_m)$$

$$\therefore \quad \rho_{m+1}|h.r_m = \frac{r_{m+1}|h}{r_m|h}$$

THE PRINCIPLES OF PROBLEMATIC INDUCTION

But
$$r_{m+1} \equiv :(r_{m+1}.R_0). \vee \ldots .(r_{m+1}.R_s). \vee \ldots (r_{m+1}.R_n)$$
$$\therefore \quad r_{m+1}|h = \sum_{s=0}^{s=n} (r_{m+1}.R_s)|h$$
$$= \sum_{s=0}^{s=n} (R_s|h)(r_{m+1}|h.R_s)$$

On the assumption that each counter drawn is replaced before the next draw,
$$r_{m+1}|h.R_s = \left(\frac{s}{n}\right)^{m+1}$$

Now
$$R_s|h = {}^nC_s \left(\frac{1}{v}\right)^s \left(1 - \frac{1}{v}\right)^{n-s} = \frac{{}^nC_s(v-1)^{n-s}}{v^n}.$$

So
$$r_{m+1}|h.R_s = \frac{\sum_{s=0}^{s=n} {}^nC_s(v-1)^{n-s} s^{m+1}}{v^n n^{m+1}}.$$

By similar reasoning,
$$r_m|h.R_s = \frac{\sum_{s=0}^{s=n} {}^nC_s(v-1)^{n-s} s^m}{v^n n^m}$$

$$\therefore \quad p_{m+1}|h.r_m = \frac{\sum_{s=0}^{s=n} {}^nC_s(v-1)^{n-s} s^{m+1}}{n \sum_{s=0}^{s=n} {}^nC_s(v-1)^{n-s} s^m}.$$

Now we can write for s^m
$$s^m = \sum_{t=1}^{t=m} A_t^m s(s-1)\ldots(s-t+1)$$

where the coefficients A_t^m are constants.

$$\therefore \quad \sum_{s=0}^{s=n} {}^nC_s(v-1)^{n-s} s^m$$
$$= \sum_{s=0}^{s=n} {}^nC_s(v-1)^{n-s} \sum_{t=1}^{t=m} A_t^m s(s-1)\ldots(s-t+1)$$

$$= \sum_{t=1}^{t=m} A_t^m \sum_{s=0}^{s=n} {}^nC_s(v-1)^{n-s} s(s-1)\ldots(s-t+1)$$

$$= \sum_{t=1}^{t=m} A_t^m \sum_{s=t}^{s=n} {}^nC_s(v-1)^{n-s} s(s-1)\ldots(s-t+1)$$

But

$${}^nC_s = \frac{n!}{s!(n-s)!}$$

and

$$s(s-1)\ldots(s-t+1) = \frac{s!}{(s-t)!}$$

$$\therefore \quad {}^nC_s s(s-1)\ldots(s-t+1)$$

$$= \frac{n!}{(n-s)!(s-t)!} = \frac{n!}{(n-t)!} \frac{(n-t)!}{(n-t-s+t)!(s-t)!}$$

$$= \frac{n!}{(n-t)!} {}^{n-t}C_{s-t}.$$

Again $(v-1)^{n-s} = (v-1)^{n-t-(s-t)}$

$$\therefore \quad \sum_{s=0}^{s=n} {}^nC_s(v-1)^{n-s} s^m$$

$$= \sum_{t=1}^{t=m} A_t^m \frac{n!}{(n-t)!} \sum_{s=t}^{s=n} {}^{n-t}C_{s-t}(v-1)^{n-t-(s-t)}$$

Put $s-t=u$.
Then when $s=t$, $u=0$; and when $s=n$, $u=n-t$.

So

$$\sum_{s=0}^{s=n} {}^{n-t}C_{s-t}(v-1)^{n-t-(s-t)} = \sum_{u=0}^{u=n-t} {}^{n-t}C_u(v-1)^{n-t-u}$$

$$= v^{n-t}$$

So

$$\sum_{s=t}^{s=n} {}^nC_s(v-1)^{n-s} s^m = \sum_{t=1}^{t=m} A_t^m \frac{n! v^{n-t}}{(n-t)!}$$

By similar reasoning

$$\sum_{s=0}^{s=n} {}^nC_s(v-1)^{n-s} s^{m+1} = \sum_{t=1}^{t=m+1} A_t^{m+1} \frac{n! v^{n-t}}{(n-t)!}.$$

Substituting these values in (I), we get finally

(I') $$p_{m+1}|h.r_m = \frac{\sum_{t=1}^{t=m+1} A_t^{m+1} \frac{v^{n-t}}{(n-t)!}}{n \sum_{t=1}^{t=m} A_t^m \frac{v^{n-t}}{(n-t)!}}$$

2. Nomic generalization

In this case we have to evaluate the probability $R_n|h.r_m$. Now

$$R_n|h.r_m = \frac{(R_n|h)(r_m|h.R_n)}{r_m|h}$$

$$= \frac{(R_n|h)(r_m|h.R_n)}{\sum_{s=0}^{s=n}(r_m.R_s)|h}$$

$$= \frac{(R_n|h)(r_m|h.R_n)}{\sum_{s=0}^{s=n}(R_s|h)(r_m|h.R_s)}$$

Now

$$R_n|h = \left(\frac{1}{v}\right)^n; \quad r_m|h.R_n = 1; \quad R_s|h = {}^nC_s\left(\frac{1}{v}\right)^s\left(1-\frac{1}{v}\right)^{n-s}$$

$$= \frac{{}^nC_s(v-1)^{n-s}}{v^n};$$

and

$$r_m|h.R_s = \left(\frac{s}{n}\right)^m$$

(II) $$\therefore R_n|h.r_m = \frac{n^m}{\sum_{s=0}^{s=n} {}^nC_s(v-1)^{n-s} s^m}$$

125

$$\text{(II}')\qquad = \frac{n^m}{n!\sum\limits_{t=1}^{t=m} A_t^m \dfrac{v^{n-t}}{(n-t)!}}.$$

NOTES

[1] We constantly use what is an immediate consequence of this, *viz.*,
$$p|qh = \frac{(p|h)(q|ph)}{(q|h)}.$$

[2] I have to thank Mr. A. E. Ingham, Fellow of Trinity, for kindly supplying me with the proof which follows.

THE PRINCIPLES OF DEMONSTRATIVE INDUCTION

Some years ago I wrote two articles in Mind on *Induction and Probability*, and, more recently, in my presidential address to the Aristotelian Society I tried to state as fully and clearly as I could the present position of the logical theory of what Mr. Johnson calls 'Problematic Induction'. In the present paper I propose to do the same for what he calls 'Demonstrative Induction.' In the former undertaking I was greatly indebted to Mr. Keynes, and in this I am even more indebted to Mr. Johnson. All my raw material is contained in his work on *Logic*, and I can claim no more than to have beaten it into a more coherent shape than that in which he left it. I think that my approach to the subject by way of the notions of Necessary and Sufficient Conditions has certain advantages, and that I have been able to make some extensions of the theory. This must be my excuse for publishing a rather long and tedious essay on a somewhat hackneyed subject which has been treated so fully and so recently by a logician of Mr. Johnson's eminence.

1. DEFINITION OF 'DEMONSTRATIVE INDUCTION'

A demonstrative induction is a mixed hypothetical syllogism of the form *Modus Ponendo Ponens* (*i.e.*, if p then q, But p, Therefore q), in which the premises are of a certain form. The major premise must be either of the form (a) If *this* S is P then *all* S is P, or (b) If *at least one* S is P then *all* S is P. In the first case the minor premise must be of the form *This* (same) S is P. In the second case the minor premise must be either of the form *This* S is P, or of the form *At least one* S is P. (It is of course obvious that the former implies the latter, whilst the latter does not imply the former.) The conclusion is always of the form *All* S is P.

We can sum this up in words as follows. The major premise must be a hypothetical proposition, in which the consequent is a universal categorical, and the antecedent is either a singular or a particular categorical of the same quality and with the same subject and predicate terms as the consequent. The minor premise must be the antecedent in the major if this be *singular*. If the antecedent in the major be *particular* the minor premise may be either this antecedent or may be a singular proposition with the same subject and predicate terms and the same quality as the antecedent. The conclusion is always the consequent in the major premise.

In the notation of *Principia Mathematica* the three forms of demonstrative induction may be symbolised as follows:

(I) $\phi a . \psi a : \supset : \phi x \supset_x \psi x$
$\phi a . \psi a$
$\therefore \quad \phi x \supset_x \psi x .$

(IIa) $(\exists x) . \phi x . \psi x : \supset : \phi x \supset_x \psi x$
$\phi a . \psi a$
$\therefore \quad \phi x \supset_x \psi x .$

(IIb) $(\exists x) . \phi x . \psi x : \supset : \phi x \supset_x \psi x$
$(\exists x) . \phi x . \psi x$
$\therefore \quad \phi x \supset_x \psi x .$

An example would be: 'If someone who sleeps in the dormitory has measles, then everyone who sleeps in the dormitory will have measles. But Jones sleeps in the dormitory and has measles. (Or, alternatively, Someone who sleeps in the dormitory has measles.) Therefore everyone who sleeps in the dormitory will have measles.' This illustrates IIa and IIb. The following would illustrate I: 'If the gas Hydrogen can be liquefied, then every gas can be liquefied. But the gas Hydrogen can be liquefied. Therefore every gas can be liquefied.' I think it is worth while to note that when we use a major premise of this form we are generally taking an extreme instance (*e.g.*, Hydrogen, because it is the lightest and most 'gassy' of all gases), and then arguing that if *even* this has a certain property all other members of the same class will *a fortiori* have it. Another example would be the premise: 'If the philosopher X can detect no fallacy in this argument no philosopher will be able to detect a fallacy in it.' We might be prepared to accept this premise on the grounds of the extreme acuteness of X. But we certainly should not be prepared to accept the premise: 'If some philosopher or other can detect no fallacy in this argument then no philosopher will be able to detect a fallacy in it.' For the philosopher Y might well rush in where X would fear to tread.

In all cases that we are likely to have to consider, the major premise of a demonstrative induction rests ultimately on a problematic induction. In all such cases it will only have a certain degree of probability. Consequently, although the conclusions of demonstrative inductions do follow of necessity from their premises, they are only probable, because one at least of the premises is only probable. It may happen that both the premises are only probable. Take, *e.g.*, Mr. Johnson's example about the atomic weight of Argon. The ultimate major premise is no doubt the proposition that if some sample of a chemical element has a certain atomic weight then all samples of that element will have that atomic weight. This is a problematic induction from an enormous number of chemical facts, and is only probable. (In fact, owing to the existence of Isotopes, it is not unconditionally true.) But one would also need the premise that Argon is a chemical element. This is again a problematic induction from a large number of chemical facts. And it is only probable.

The argument about Argon, when fully stated, would take the following form: (i) If some sample of a chemical element has a certain atomic weight, then all samples of that element will have that atomic weight. But Argon is a chemical element. Therefore if some sample of Argon has a certain atomic weight W all samples of Argon will have the atomic weight W. (This is an ordinary syllogism.) (ii) Therefore if some specimen of Argon has the atomic weight 40 all specimens of Argon will have the atomic weight 40. (This is a conclusion drawn by the Applicative Principle.) (iii) This specimen of Argon has atomic weight 40. Therefore all specimens of Argon will have atomic weight 40. (This is the demonstrative induction.) The empirical premises are three, *viz.*, the original generalization about chemical elements, the proposition that Argon is an element, and the proposition that the atomic weight of this specimen of Argon is 40.

Now much the most important major premises for demonstrative inductions are provided by causal laws. It will therefore be necessary for us to consider next the question of Causal Laws.

2. CAUSAL LAWS

The word 'cause' is used very ambiguously in ordinary life and even in science. Sometimes it means a necessary, but it may be insufficient, condition (*e.g.*, 'sparks cause fires'). Sometimes it means a sufficient, but it may be more than sufficient, condition or set of conditions (*e.g.*, 'Falling from a cliff causes concussion'). Sometimes it means a set of conditions which are severally necessary and jointly sufficient. But, in any interpretation, it involves one or both of the notions of 'necessary' and 'sufficient' condition. It is therefore essential to begin by defining these notions and proving the most important general propositions that are true about them.

There is one other preliminary remark to be made. There are two different types of causal law, a cruder and a more advanced. The cruder type merely asserts connexions between *determinable* characteristics. It just says that whenever such and such determinable characteristics are present such and such another determinable characteristic will be present. An example would be the law that cloven-footed animals chew the cud, or that rise of temperature causes bodies to expand. I shall call such laws '*Laws of Conjunction of Determinables*'. The more advanced type of law considers the determinate values of conjoined determinables. It gives a formula from which the determinate values of the effect-determinables can be calculated for every possible set of determinate values of the cause-determinables. An example would be the law for gases that $P = RT/V$. I will call such laws '*Laws of Correlated Variation of Determinates*'. In the early stages of any science the laws are of the first kind, and in many sciences they have never got beyond this stage, *e.g.*, in biology and psychology. But the ideal of every science is to advance from laws of the first kind to laws of the second kind. Now Mill's Methods of Agreement, Difference, and the Joint Method, are wholly concerned with the establishment of laws of conjunction of determinables. His Method of Concomitant Variations *ought* to have been concerned with the establishment of laws of correlated variation of determinates. But, since he talks of it as simply a weaker form of the Method of Difference, which we have to put up with when circumstances will not allows us to use that method, it is plain that he did not view it in this light. On the other hand, Mr. Johnson's Methods are definitely concerned with laws of correlated variation. They presuppose that laws of conjunction of determinables have already been established.

The order which I shall follow henceforth is this: (i) I shall deal with the notion of necessary and sufficient conditions wholly in terms of determinables. I shall then state Mill's Methods in strict logical form and show what each of them would really prove. (ii) I shall then pass to the notion of correlated variation of determinates, and explain Mr. Johnson's methods.

3. NECESSARY AND SUFFICIENT CONDITIONS

i. *Notation*

The letters E, and C_1, C_2, etc., are to stand for determinable characteristics. I shall use C's to denote determining factors and E's to denote determined factors.

ii. *Definitions*

'C is a *sufficient condition* (S.C.) of E' means 'Everything that has C has E' (1).

'C is a *necessary condition* (N.C.) of E' means 'Everything that has E has C' (2).

'$C_1 \ldots C_n$ is a *smallest sufficient condition* (S.S.C.) of E' means that '$C_1 \ldots C_n$ is a S.C. of E, and no selection of factors from $C_1 \ldots C_n$ is a S.C. of E' (3).

'$C_1 \ldots C_m$ is a *greatest necessary condition* (G.N.C.) of E' means that 'C_1 and C_2 and $\ldots C_m$ are each a N.C. of E, but nothing outside this set is a N.C. of E' (4).

'$C_1 \ldots C_n$ are *severally necessary and jointly sufficient* to produce E' means that '$C_1 \ldots C_n$ is both a S.S.C. and a G.N.C. of E' (5).

(*N.B.* I have represented the effect-determinable by the single letter E. This is not meant to imply that it really consists of a single determinable characteristic. In general, it will be complex, like the cause-determinable, and will be of the form $E_1 \ldots E_m$. But in the propositions which I am going to prove in the next few pages the complexity of the effect-determinable is irrelevant, and so it is harmless and convenient to denote it by a single letter. Later on I shall prove a few propositions in which it is necessary to take explicit account of its internal complexity.)

iii. *Postulates*

(1) It is assumed that all the C-factors are capable of independent presence or absence. This involves (a) that none of them is either a conjunction or alternation of any of the others. (*E.g.*, C_3 must not be the conjunctive characteristic C_1-and-C_2. Nor may it be the alternative characteristic C_1-or-C_2.) Again (b) no two of them must be related as red is to colour (for then the first could not occur without the second), or as red is to green (for then the two could not occur together). It is also necessary to assume that all combinations are *causally* possible. For otherwise we might have the two causal laws 'Everything that has C_1C_2 has C_3' and 'Everything that has C_3 has E'. In that case both C_1C_2 and C_3 would have to be counted as S.C.'s of E, since the law 'Everything that has C_1C_2 has E' would follow as a logical consequence of these two other laws. This would obviously be inconvenient; we want to confine our attention to *ultimate* causal laws. Our present postulate may be summed up in the proposition that, if there be n cause-factors, it is assumed that all the 2^n-1 possible selections (including all taken together) are both logically and causally possible. This may be called the 'Postulate of *Conjunctive Independence*'.

(2) It is further assumed that every occurrence of any determinable characteristic E has a S.S.C. This means that, whenever the characteristic E occurs, there is some set of characteristics (not necessarily the same in each case) such that the presence of this set in any substance carries with it the presence of E, whilst the presence of any selection from this set is consistent with the absence of E. This is the form which the Law of Universal Causation takes for the present purpose. We will call it 'The Postulate of *Smallest Sufficient Conditions*.'

iv. *Propositions*

(1) 'If C be a S.C. of E, then any set of conditions which contains C as a factor will also be a S.C. of E.'

Let such a set of conditions be denoted by CX.
Then: (a) All that has CX has C.
      (b) All that has C has E. (Df. 1.)
Therefore all that has CX has E.
Therefore CX is a S.C. of E. (Df. 1.)   Q.E.D.

(2) 'If $C_1 \ldots C_m$ be a N.C. of E, then any set of conditions contained in $C_1 \ldots C_m$ will also be a N.C. of E.'

Consider, *e.g.*, the selection C_1C_2.

Then: (a) All that has $C_1 \ldots C_m$ has C_1C_2.
     (b) All that has E has $C_1 \ldots C_m$. (Df. 2.)

Therefore all that has E has C_1C_2.

Therefore C_1C_2 is a N.C. of E. (Df. 2.)   Q.E.D.

(3) 'Any S.C. of E must contain all the N.C.'s of E.'

Let X be a S.C. of E, and let Y be a N.C. of E.

Then: (a) All that has X has E. (Df. 1.)
     (b) All that has E has Y. (Df. 2.)

Therefore all that has X has Y.

But all the C's are capable of independent presence or absence. (Postulate 1.) Hence this can be true only if X be of the form YZ.

Therefore any S.C. of E. must contain as factors every N.C. of E, if E has any N.C.'s.   Q.E.D.

(4) 'E cannot have more than one G.N.C.'

Let $C_1 \ldots C_m$ be a G.N.C. of E. Then this set (a) contains *nothing but* N.C.'s of E (Prop. 2); and (b) contains *all* the N.C.'s of E. (Df. 4.)

Now any alternative set must either (a) contain some factor which is not contained in this one; or (b) contain no factor which is not contained in this one. In the first case it will contain some factors which are not N.C.'s of E. Therefore such a set could not be a G.N.C. of E. In the second case this set either coincides with $C_1 \ldots C_m$ or is a selection from $C_1 \ldots C_m$. On the first alternative it does not differ from $C_1 \ldots C_m$. On the second alternative it does not contain *all* the N.C.'s of E.

Therefore it could not be a G.N.C. of E.

Therefore E cannot have more than one G.N.C.   Q.E.D.

(5) 'E can have a plurality of S.S.C.'s. These may be either entirely independent of each other, or they may partially overlap; but one cannot be wholly contained in the other.'

Take, *e.g.*, C_1C_2 and $C_3C_4C_5$.

To say that C_1C_2 is a S.S.C. of E is to say that everything which has $C_1 C_2$ has E; whilst C_1 can occur without E, and C_2 can occur without E. (Df. 3.)

To say that $C_3C_4C_5$ is a S.S.C. of E is to say that everything which has

$C_3C_4C_5$ has E; whilst C_3C_4 can occur without E, and C_4C_5 can occur without E, and C_5C_3 can occur without E. (Df. 3.)

It is evident that the two sets of statements are logically independent of each other and can both be true.

Now take C_1C_2 and C_2C_3.

We have already stated what is meant by saying that C_1C_2 is a S.S.C. of E. To say that C_2C_3 is a S.S.C. of E means that everything which has C_2C_3 has E; whilst C_2 can occur without E, and C_3 can occur without E. If the two sets of statements be compared it will be seen that they are quite compatible with each other.

But it would be impossible, *e.g.*, for C_1C_2 and C_1 to be both of them S.S.C.'s of E. For, if C_1C_2 were a S.S.C., it would follow from Df. 3 that C_1 would not be a S.C. at all. Q.E.D.

(6) 'Any factor which is common to all the S.S.C.'s of E is a N.C. of E.'

Let S_1, S_2, and S_3 be *all* the S.S.C.'s of E. And let C be a factor common to all of them.

Since every occurrence of E has a S.S.C. (Postulate 2), everything that has E has either S_1 or S_2 or S_3.

But everything that has S_1 has C, and everything that has S_2 has C, and everything that has S_3 has C.

Therefore everything that has E has C.

Therefore C is a N.C. of E. (Df. 2.) Q.E.D.

(7) 'If E has *only one* S.S.C., it has also a G.N.C., and these two are identical. And so this set is severally necessary and jointly sufficient to produce E.'

By Prop. 4 there cannot be more than one G.N.C. of E.

By Prop. 3 the S.S.C. of E must contain the G.N.C. of E.

By Prop. 6 any factor that is common to all the S.S.C.'s of E must be a N.C. of E. Now, since in the present case there is *only one* S.S.C. of E, *every* factor in it is common to all the S.S.C.'s of E.

Therefore every factor in the S.S.C. of E is a N.C. of E.

But we have already shown that every N.C. of E must be a factor in the S.S.C. of E.

Therefore the S.S.C. and the G.N.C. of E coincide.

Therefore this set of factors is severally necessary and jointly sufficient to produce E. Q.E.D.

(8) 'If C be a S.C. of E_1 and also a S.C. of E_2, then it will also be a S.C. of E_1E_2. And the converse of this holds also.'

The hypothesis is equivalent to the two propositions:
        All that has C has E_1; and
        All that has C has E_2. (Df. 1.)

Now these are together equivalent to the proposition: 'All that has C has E_1E_2.' And this is equivalent to the proposition: 'C is a S.C. of E_1E_2.' (Df. 1.)   Q.E.D.

(9) 'If C be a N.C. of either E_1 or E_2, then it is a N.C. of E_1E_2.'

If C be a N.C. of E_1 it follows from Df. 2 that all that has E_1 has C. But all that has E_1E_2 has E_1.

Therefore all that has E_1E_2 has C.

Therefore, by Df. 2, C is a N.C. of E_1E_2.

In exactly the same way it can be shown that, if C be a N.C. of E_2, it will be a N.C. of E_1E_2.

Therefore, if C be a N.C. either of E_1 or of E_2, it will be a N.C. of E_1E_2.
        Q.E.D.

(10) 'The converse of (9) is false. It is possible for C to be a N.C. of E_1E_2 without its being a N.C. of E_1 or a N.C. of E_2.'

If C be a N.C. of E_1E_2, then all that has E_1E_2 has C. (Df. 2.)

But this is quite compatible with there being some things which have E_1 without having C, or with there being some things which have E_2 without having C. (*E.g.*, all things that are black and human have woolly hair. But there are black things and there are human things which do not have woolly hair.)

So the truth of the proposition that C is a N.C. of E_1E_2 is compatible with the falsity of either or both the propositions that C is a N.C. of E_1 and that C is a N.C. of E_2.   Q.E.D.

(11) 'If C_1C_2 be a S.C. of *each* of the effect-factors $E_1, E_2, ..., E_n$, and if it be a S.S.C. of *at least one* of them, then it will be a S.S.C. of the complex effect $E_1 ... E_n$.'

From Prop. 8 it follows at once that C_1C_2 will be a S.C. of $E_1 ... E_n$. It is therefore only necessary to show that it will be a S.S.C.

Let us suppose, *e.g.*, that C_1C_2 is a S.S.C. of the factor E_1. Then, from Df. 3, it follows that C_1 is not a S.C. of E_1 and that C_2 is not a S.C. of E_1.

Now suppose, if possible, that C_1C_2 is not a S.S.C. of $E_1 ... E_n$. We know that it is a S.C. of $E_1 \ldots E_n$. If it be not a S.S.C., then either C_1 or C_2 must be a S.C. of $E_1 \ldots E_2$. (Df. 3.) But, if so, then either C_1 or C_2 must be a S.C. of E_1. (Prop. 8.) But we have seen above that neither C_1 nor C_2 can be a S.C. of E_1.

Hence the supposition that $C_1 C_2$ is not a S.S.C. of $E_1 \ldots E_n$ is impossible. Q.E.D.

(12) 'The converse of (11) is false. If $C_1 C_2$ be a S.S.C. of $E_1 \ldots E_n$, it will indeed be a S.C. of each of the factors; but it need not be a S.S.C. of any of the factors.'

This is obvious. *E.g.*, C_1 might be sufficient to produce E_1, though nothing less than $C_1 C_2$ was sufficient to produce $E_1 \ldots E_n$.

4. THE POPULAR-SCIENTIFIC NOTION OF 'CAUSE' AND 'EFFECT'

The notions which we have been defining and discussing above are those which emerge from the looser notions of 'cause' and 'effect', which are current in daily life and the sciences, when we try to make them precise and susceptible of logical manipulation. There are, however, certain points which must be cleared up before the exact relation between the logical and the popular-scientific notions can be seen.

i. *The time-factor*

It might well be objected that the notion of temporal succession is an essential factor in the common view of cause and effect, and that this has disappeared in our account of necessary and sufficient conditions. The effect is conceived as something that begins at the same moment as the cause ends. And without this temporal distinction it would be impossible to distinguish effect from cause. All this is perfectly true, and it would be of great importance to make it quite explicit if one were dealing with the metaphysics, as distinct from the mere logical manipulation, of causation. But for the present purpose it may be met by the following remark about our notation. We must think of some at least of our C's being really of the complex form 'being characterised by $\mathfrak{C}$ up to the moment t', and of some of our E's as being really of the complex form 'beginning to be characterised by $\mathfrak{E}$ at the moment t'.

ii. *Transeunt causation*

A second highly plausible objection would be the following. In our exposition of necessary and sufficient conditions we have always talked of a single continuant, and have supposed that the effect-characteristics and the cause-characteristics occur in the same continuant. But in fact most causation is transeunt, *i.e.*, the cause-event takes place in one continuant and the effect-event in another. This, again, is perfectly true, and very important in any attempt at an analysis of causation for metaphysical purposes. The usual kind of causal law does in fact take roughly the following form: 'if a continuant having the properties P is in the state S_1 at a moment t and it then comes into the relation R to a continuant which has the properties P' and is in the state S'_1, the former continuant will begin to be in the state S_2 and the latter in the state S'_2.' *E.g.*, 'If a hard massive body moving in a certain direction and with a certain velocity at a certain moment comes at that moment into contact with a soft inelastic body at rest, the motion of the former body will begin to change and a dint will develop in the latter body.'

For mere purposes of logical manipulation, however, all this can be symbolised as changes in the characteristics of the first continuant. We shall have to remember that some of our C's and some of our E's stand for relational properties of a very complex kind, involving relations to other continuants. Thus, in the example one of our C's will be the characteristic of 'Coming into contact at t with a soft inelastic resting body'. And one of our E's will be the characteristic of 'Having been in contact at t with the same body beginning to develop a dint'. All this is purely a matter of verbal and notational convenience. It has no philosophical significance. But it is harmless so long as we remember that our innocent-looking C's and E's stand, not just for simple qualities but for extremely complex relational properties of the various kinds described above.

iii. *Negative factors*

It must be clearly understood that some of the C's and some of the E's may stand for negative characteristics, *i.e.*, for the absence of certain positive characteristics. Negative conditions may be just as important as positive ones. *E.g.*, there is no general law about the effect of heat on oxygen. If the oxygen be free from contact with other gases it merely expands when heated. If it be mixed with a sufficient proportion of hydrogen it explodes. Thus the negative condition 'in absence of hydrogen' is an essential factor when the effect to be considered is the expansion of oxygen.

5. PLURALITY OF CAUSES AND EFFECTS

i. *Total cause and total effect*

Before we can discuss whether plurality of causes or of effects is logically possible we must define the notions of 'total cause' and 'total effect'. The definition is as follows:

'$C_1 \ldots C_n$ stands to $E_1 \ldots E_m$ in the relation of *total cause to total effect*' means that '$C_1 \ldots C_n$ is a S.S.C. of $E_1 \ldots E_m$, and it is not a S.C. of any characteristic outside the set $E_1 \ldots E_m$.' (Df. 6.)

It will be seen that this definition is equivalent to the conjunction of the following three propositions, one of which is affirmative and the other two negative:

(a) Any occurrence of $C_1 \ldots C_n$ is also an occurrence of $E_1 \ldots E_m$.

(b) There is no selection of factors from $C_1 \ldots C_n$ such that every occurrence of it is also an occurrence of $E_1 \ldots E_m$.

(c) There is no factor outside $E_1 \ldots E_m$ such that every occurrence of $C_1 \ldots C_n$ is also an occurrence of it.

ii. *Plurality of causes*

With this definition it is logically possible for *several* different sets of factors to stand to *one and the same* set of factors in the relation of total cause to total effect. For we have proved in Prop. 5 that one and the same E can have a plurality of different S.S.C.'s. We also showed there that the various S.S.C.'s may either have no factor in common or may partially overlap, but that one cannot be wholly included in another. We also showed in Prop. 6 that any factor which is common to all possible S.S.C.'s of a given E is a N.C. of that E. It is, of course, quite possible for an effect to have *no* necessary conditions. For if it has two S.S.C.'s which have no factor in common, it cannot possibly have a N.C. On the other hand

(Prop. 7), if an effect has only one S.S.C. this is also the G.N.C. of the effect. So, when there is no plurality of causes, the total cause of a given total effect is a set of factors which are severally necessary and jointly sufficient to produce the effect.

Thus our definitions allow the possibility of a plurality of *total causes* for one and the same *total effect*. Whether there actually is such plurality in nature, or whether the appearance of it is always due to our partial ignorance or inadequate analysis, is a question into which I shall not enter here. Of course, even if a given total effect does have a plurality of total causes, *each particular occurrence* of this total effect will be determined by the occurrence of one and only one of these total causes. The plurality will show itself in the fact that some occurrences of the total effect will be determined by occurrences of one of the total causes, whilst other occurrences of the total effect will be determined by occurrences of another of the total causes.

iii. *Plurality of effects*

It is plain from Df. 6 that a given total cause could not have more than one total effect. Thus plurality of total effects is ruled out by our definitions.

6. FORMAL STATEMENT OF MILL'S METHODS

We are now in a position to deal with Mill's Methods of Agreement and Difference. Mill never clearly defined what he meant by 'cause' or by 'effect,' and he never clearly stated what suppressed premises, if any, were needed by his Methods. We shall now be able to see exactly in what sense 'cause' and 'effect' are used in each application of each Method; what assumptions are tacitly made; and what bearing the question of 'plurality of causes' has on the validity of each application of each Method. Mill made two applications of each Method, *viz.*, to find 'the effect of a given cause' and to find 'the cause of a given effect'. We have therefore in all four cases to consider:

i. *Method of agreement*

(a) *To find the 'effect' of* A. The premises are:

All ABC is *abc*; and
All ADE is *ade*.

The argument should then run as follows:

A is not a S.C. of *bc*; for in the second case A occurs without *bc*. It is assumed that A is a S.C. of *something* in *abc*. Therefore it must be a sufficient condition of *a*.

Thus, the suppressed premise is that A is a S.C. of something or other in *abc*. And the sense in which it is proved that the effect of A is *a* is that it is *a* of which A is a S.C.

(b) *To find the 'cause' of a.*

The premises are as before.

The argument should run as follows:

From the two premises it follows that both ABC and ADE are S.C.'s of *a*. But every S.C. of *a* must contain all the N.C.'s of *a*. (Prop. 3.)

Therefore, if *a* has a N.C. at all, it must be or be contained in the common part of the two S.C.'s of *a*.

But the only common part is A.

Therefore, if *a* has a N.C. at all, either A itself or some part of A must be a N.C. of *a*.

Thus the sense in which it is proved that the cause of *a* is A or some part of A is that if *a* has a N.C. at all then it is A or some part of A which is its N.C.

Mill's contention that, in this application, the Method of Agreement is rendered uncertain by the possibility of Plurality of Causes is true, and has the following meaning. If it be admitted that *a* may have more than one S.S.C. it is possible that it may have no N.C. at all. In fact, this will be the case if there is no factor common to all its S.S.C.'s. Thus, we cannot draw the categorical conclusion that the N.C. of *a* is or is contained in A unless we are given the additional premise: '*a* has either only one S.S.C., or, if it has several, there is a factor common to all of them.'

ii. *Method of difference*

(a) *To find the 'effect' of* A. The premises are:

All ABC is *abc*; and
All (non-A) BC is (non-*a*) *bc*.

The argument should run as follows:
A is not a N.C. of *bc*, for in the second case *bc* occurs without A.

THE PRINCIPLES OF DEMONSTRATIVE INDUCTION

It is assumed that A is a N.C. of *something* in *abc*.
Therefore A must be a N.C. of *a*.

Thus the suppressed premise is that A is a N.C. of *something* in *abc*. And the sense in which it is proved that the effect of A is *a* is that it is *a* of which A is a N.C.

(b) *To find the 'cause' of a.*

The premises are as before.

The argument should run as follows:

It follows from the second premise that All (non-A) BC is non-*a*.

Therefore, by contraposition, All *a* is non-[(non-A)BC].

Therefore, All *a* is either A or non-(BC).

Therefore, All *a* which is BC is A.

This may be stated in the form: 'In presence of BC, A is necessary to produce *a*.'

Now, the first premise could be put in the form: 'In presence of BC, A is sufficient to produce *a*'.

Combining these, we reach the final conclusion: 'In presence of BC, A is necessary and sufficient to produce *a*'.

We have no right to conclude that A would be either necessary or sufficient in the absence of BC. In the presence of a suitable mixture of hydrogen and oxygen a spark is both necessary and sufficient to produce an explosion with the formation of water. But it is not sufficient in the absence of either of the two gases. Again, when a person is in good general health, prolonged and concentrated exposure to infection is necessary and sufficient to give him a cold. But when he is in bad general health it is not necessary that the exposure should be either prolonged or concentrated.

Thus Mill has no right to draw the unqualified conclusion that A is the cause of *a*, either in the sense of necessary or in the sense of sufficient condition. But he is justified in concluding that, in presence of BC, A is the cause of *a*, in the sense of being necessary and sufficient to produce *a*.

iii. *The joint method*

Mill's Joint Method is suggested as a method by which we may find the 'cause' of *a* in cases where the Method of Difference cannot be used, and where the Method of Agreement is rendered untrustworthy by the possibility of Plurality of Causes.

It consists of two parts. The first is an ordinary application of the Method of Agreement. From this we reach the conclusion that, unless *a* has a plurality of S.S.C.'s with no factor common to all of them, A or some part of A is a N.C. of *a*. But, owing to the possibility of plurality of causes, it remains possible that A may be irrelevant to *a*. It may be, *e.g.*, that BC is a S.S.C. of *a* in the first case, and that DE is a S.S.C. of *a* in the second case, and therefore that *a* has no N.C. at all. The second part of the Joint Method is supposed to state conditions under which this possibility might be rejected. It is as follows. We are to look for a pair of instances which agree in *no* respect, positive or negative, except that A and *a* are absent from both of them. It is alleged by Mill that, if we find such a pair of instances, we can conclude with certainty that the 'cause' of *a* is A.

It is, of course, quite plain that, even if the method were logically unimpeachable, it would be perfectly useless in practice. Any pair of instances that we could possibly find would agree in innumerable *negative* characteristics beside the absence of A and the absence of *a*. But is the argument logically sound even if premises of the required kind could be found?

It would run as follows. Since our two instances are to agree in *no* respect, positive or negative, except the absence of A and of *a*, BC cannot be absent in both of them. Therefore BC must be present in one of them. But *a* is absent in both of them. Therefore, in one of them BC is present without *a* being present. Therefore BC cannot be a S.C. of *a*. But, from the first part of the method, we know that ABC is a S.C. of *a*. A precisely similar argument would show that DE cannot be a S.C. of *a*. And, from the first part of the method, we know that ADE is a S.C. of *a*. Mill thinks that we can conclude that A is a N.C. of *a*. This, however, is a mistake. All that we can conclude is that, *in presence of BC or DE*, A is a N.C. of *a*. It remains quite possible that there is another S.S.C. of *a*, *e.g.*, XYZ, which does not contain A at all. And, in that case, A could not be a N.C., without qualification, of *a*. *E.g.*, a certain kind of soil, when treated with lime, always yields good crops; and, when lime is absent, good crops are absent on this soil. This proves that the presence of lime is a necessary condition for getting good crops *with this kind of soil*. But it does not prove that the presence of lime is a necessary condition, without qualification, for getting good crops. With other kinds of soil it might be unnecessary or positively harmful.

THE PRINCIPLES OF DEMONSTRATIVE INDUCTION

There is, however, a perfectly sensible method of argument, which is not Mill's, but which might fairly be called the *Joint Method*. The first part of it would be to take large number of sets of characteristics, such that each set contains A and that in other respects they are as unlike each other as possible. One would try to arrange that A should be the only characteristic common to *all* of them, though it might be impossible to arrange that any *two* of them had only A in common. Suppose it were found that every occurrence of each of these sets was also an occurrence of a. Then there would be a strong presumption, though never a rigid proof, that A was a S.C. of a. The alternative would be that a had an enormous number of alternative S.S.C.'s. The second part of the method would be to take a large number of sets of characteristics, such that each set *lacks* A, and that in other respects they are as unlike each other as possible. One would try to arrange that non-A should be the only characteristic common to *all* of them, though it might be impossible to arrange that any *two* of them had only non-A in common. Suppose it were found that every occurrence of each of these sets was also characterised by the *absence* of a. Then there would be a strong presumption, though never a rigid proof, that non-A was a S.C. of non-a. It would then follow by contraposition that A was a N.C. of a. Thus the combination of the two sets of observations would make it probable that A is a necessary and sufficient condition of a. The argument is, of course, greatly strengthened if the characteristics other than A and a which occur among the sets of the first series are, as nearly as may be, the same as the characteristics other than non-A and non-a which occur among the sets of the second series. Thus, as Mr. Johnson has pointed out, the various sets of the same series should differ as much as possible in all respects except the one under investigation; whilst the two series, as wholes, should agree as much as possible in all respects except the one under investigation. A good example would be provided by the empirical arguments which lead to the conclusion that the property of having an asymmetrical molecular structure is a necessary and sufficient condition of the property of rotating the plane of polarisation of plane-polarised light. As Mill's own Joint Method is both useless and invalid, the name of 'Joint Method' might be reserved in future for the above important and legitimate, though not absolutely conclusive, type of inductive argument.

7. LAWS OF CORRELATED VARIATION OF DETERMINATES

We have now completed our account of the arguments by which one attempts to establish laws of *Conjunction of Determinables*. Suppose that we have thus rendered it highly probable that $C_1 \ldots C_n$ is a S.S.C. of E, where E may itself be a complex characteristic of the form $E_1 \ldots E_m$. We now want to go further and to consider the connexion between various determinate values of $C_1 \ldots C_n$, on the one hand, and various determinate values of E, on the other. This is what Mr. Johnson seeks to formulate in his inductive methods. For this purpose we need some further postulates in addition to those which we used in the theory of necessary and sufficient conditions. We will begin by stating and commenting on these postulates.

Postulates

(3) If C be a S.S.C. of E, and if there is at least one instance in which a certain determinate value c of C is accompanied by a certain determinate value e of E, then in *every* instance in which C has the value c E will have the value e. (We will call this *Postulate 3*, as we have already had two postulates.)

I will now make some comments on this postulate. (a) The converse of it is not assumed to hold. Our postulate states that c cannot be accompanied in some instances by e and in other instances by e'. But it does not deny that e may be accompanied in some cases by c and in others by c'. The point will be made clear by an example. Let E be the time of vibration of a compass-needle free to vibrate about its point of suspension in a magnetic field. Then the S.S.C. of E is a conjunction of three factors, *viz.*, the moment of inertia of the needle, its magnetic moment, and the intensity of the magnetic field. Call these three factors C_1, C_2, and C_3 respectively. Then the causal formula is in fact $E = 2\pi \sqrt{C_1/C_2 C_3}$. It is plain that, if determinate values of C_1, C_2, and C_3 be taken, any repetition of them all will involve a repetition of the original value of E. But the original value of E might occur when the values of C_1, C_2, and C_3 were different from their original values, provided the new values were suitably related among themselves.

(b) It will be noticed that the postulate is of the form required for the major premise of a demonstrative induction. For it is a hypothetical

THE PRINCIPLES OF DEMONSTRATIVE INDUCTION

proposition in which the consequent is a universal categorical, and the antecedent is a particular categorical of the same quality and with the same subject and predicate as the consequent.

(c) In virtue of this postulate we can talk of *the* value of E which corresponds to a given value of C. But we cannot talk of *the* value of C which corresponds to a given value of E, since there may be several such values. Thus the postulate may be said to deny the possibility of a plurality of determinate total effects to a given determinate total cause, but to allow of a plurality of determinate total causes to a given determinate total effect. I propose to call this postulate the 'Postulate of the *Uniqueness of the Determinate Total Effect*'.

(d) It must be clearly understood that, although in stating the postulate the single letters C and E have been used, they are meant to cover the case of conjunctions of factors, such as $C_1 \ldots C_n$ and $E_1 \ldots E_m$. In such cases the determinate c will represent the conjunction of a certain determinate value of C_1 with a certain determinate value of C_2 with … a certain determinate value of C_n. And similar remarks apply, *mutatis mutandis*, to E. Thus we shall have a different determinate value of C if we have a different determinate value of *at least one* of the determinables $C_1 \ldots C_n$, even though we have the same determinate values as before for all the other C-factors. And similar remarks apply, *mutatis mutandis*, to variations in the determinate value of E.

(4) This brings us to the fourth postulate. It runs as follows. If a total cause or a total effect be a conjunction of several determinables it is assumed that no determinate value of any of these factors either entails or excludes any determinate value of any of the other factors in this total cause or total effect. This may be called the 'Postulate of *Variational Independence*'. It should be compared with Postulate (1), which we called the postulate of *Conjunctive Independence*.

Now suppose that E is a conjunction of the determinables $E_1 \ldots E_m$. Let there be μ_1 determinates under E_1, μ_2 determinates under $E_2, \ldots$ and μ_m determinates under E_m. It follows from the Postulate of Variational Independence that the total number of different determinate values of E will be $\mu_1 \mu_2 \ldots \mu_m$. Let us call this the '*Range of Variation*' of E. Now it follows at once from the Postulate of the Uniqueness of the Determinate Total Effect that, if C be a S.S.C. of E, the range of variation of C cannot be narrower than the range of variation of E, though it may be wider. For to every different determinate value of E there must correspond a different determinate value of C, whilst several different determinate values of C may correspond to one and the same determinate value of E. Suppose that C is a conjunction of the determinables $C_1 \ldots C_n$. Let there be v_1 determinates under C_1, v_2 determinates under $C_2 \ldots$ and v_n determinates under C_n. Then the range of variation of C is $v_1 v_2 \ldots v_n$. And the principle which we have just proved is that $v_1 v_2 \ldots v_n \geqslant \mu_1 \mu_2 \ldots \mu_m$.

Now two different cases are possible. (a) Every determinable in E may have only a finite number of determinates under it. This alternative leads to nothing of great interest. (b) At least one of the determinables in E may have an infinite number of determinates under it. If so, the range of variation of E will be infinite. Consequently the range of variation of C must be infinite. But this will be secured if and only if at least one of the determinables in C has an infinite number of determinates under it. So we reach the general principle that if there is at least one factor in a total effect which has an infinite number of determinates under it then there must be at least one factor in any S.S.C. of this effect which has an infinite number of determinates under it.

We can now go rather further into detail by using the elements of Cantor's theory of transfinite cardinals. (a) Even if *all* the determinables in a total effect should have an infinite number of determinates under them it will be sufficient that *at least one* of the determinables in the total cause should have an infinite number of determinates under it. For the number of determinables in the total effect is assumed to be finite. Consequently the range of variation of the total effect will be an infinite cardinal raised to a finite power, even in the case supposed. Now it is known that any finite power of an infinite cardinal is equal to that infinite cardinal. Therefore it is enough, even in the case supposed, that at least one of the determinables in the total cause should have an infinite number of determinates under it. We can sum up our results in the form: 'If *at least one* factor in the total effect has an infinite number of determinates under it it is *necessary* that at least one factor in the total cause should have an infinite number of determinates under it; and even if *all* the factors in the total effect have an infinite number of determinates under them it is *sufficient* that at least one of the factors in the total cause should have an infinite number of determinates under it'. (b) If the number of determinates under one of the determinables in E be infinite there are still two possible alternatives. In the first place the series of determinates may merely be 'compact', *i.e.*, it may merely be the case that there is a determinate of the series between any pair of determinates of the series. If so, it has the same cardinal number as the series of finite integers, *viz.* $\aleph_0$ the smallest of the transfinite cardinals. On the other hand, the series of determinates under this determinable may be 'continuous' in the technical sense, as the points on a straight line are supposed to be. If so, it has the same cardinal number as the series of real numbers, *viz.*, $2^{\aleph_0}$. Now it is known that $2^{\aleph_0}$ is greater than $\aleph_0$. We can therefore enunciate the following general principle: 'If any of the determinables in a total effect has under it a series of determinates which is strictly "continuous" then at least one of the determinables in the total cause must have under it a series of determinates which is not merely "compact" but is strictly "continuous".'

Before leaving this subject there is one final question that might be raised. Is it possible that one or more of the determinables in a *total cause* should have an *infinite* number of determinates under it whilst all the determinables in the *total effect* have only a *finite* number of determinates under them? There is certainly nothing in any of our postulates to rule out this possibility. It would be realised if, *e.g.*, the following state of affairs existed. Suppose that C is a total cause and that E is its total effect. Suppose that E has a finite number of determinate values e_1, e_2, etc. Suppose that the determinate values of C form a compact or a continuous series. And suppose finally that c_0 and every value of C between c_0 and c_1 determines the value e_1 of E, that c_1 and every value of C between c_1 and c_2 determines the value e_2 of E, and so on. I do not see anything impossible in a law of this kind, though I do not know of any quite convincing example of such laws. The following would be at least a plausible example. Suppose we take the three possible states of a chemical substance, such as water, *viz.*, the solid, the liquid, and the gaseous, as three determinates under a determinable. And suppose we say that this determinable is a total effect of which the two determinables of pressure and temperature constitute a total cause. Keep the pressure fixed at 76 cm of mercury, and imagine the temperature to be varied continuously. Then every determinate value up to a temperature of zero on the centigrade scale determines the solid state, every determinate value from zero up to 100° determines the liquid state, and every determinate value above 100° determines the gaseous state.

I have said that the above is a *plausible* example of a case in which the same determinate total effect has an infinite plurality of different possible determinate total causes. But, when it is more carefully inspected, it can be seen not to be a *real* example. The fact is that we have not got here either the genuine total cause or the genuine total effect. The real total cause is a conjunction of three factors, *viz.*, the pressure, P, the total mass of the substance, M, and the quantity of heat contained in the substance, H. The real total effect is a conjunction of four factors; *viz.*, S, the amount of the substance in the solid state; L, the amount of the substance in the liquid state; G, the amount of the substance in the gaseous state; and T, the temperature of the substance. Our law of the conjunction of determinables is then that PMH is a S.S.C. of SLGT. Suppose that at the beginning of the experiment all the water is in the solid form, and is at a temperature below freezing-point. We will keep the determinate values of P and M constant throughout the experiment at the values p and m. And we will continuously increase H. At first L and G will have the values 0, and S will have the value m. As H is increased these values will remain constant, but T will continuously increase. This will go on till T reaches the melting-point of ice at the pressure p. If we now further increase H the values of S and L will begin to change continuously, whilst the value of T will remain at the melting-point of ice under the pressure p. The value of S will steadily diminish and that of L will steadily increase until we reach a stage at which the value of S is 0 and the value of L is m; *i.e.*, all the water will now be in the liquid state at the temperature of melting ice under the pressure p. If we still go on increasing the value of H the values of T will now start to increase steadily, and this will go on till the liquid water reaches the boiling-point under the given pressure. If H be still increased after this point we shall have the values of L and G changing, whilst T remains constant. This stage will go on as we increase H until all the water is converted into steam at the temperature of boiling water under the pressure p. At this stage S and L will have the values 0, whilst G will have the value m. If more heat be now put in, S, L, and G will henceforth keep constant at 0, 0, and m, respectively, and T will steadily rise.

We see then that at every stage *some* factor in the *total* effect is varying continuously as the factor H in the total cause varies continuously, although other factors in the total effect may at the same time be keeping constant in value. Thus the *total effect* changes continuously in value throughout the whole process, and to each determinate value of it there corresponds one and only one value of that factor in the total cause which is being continuously varied while the remaining cause-factors are kept constant. It is possible that, whenever it seems that a continuous set of different values of a total cause all determine the same value of a total effect, this is always due to our not having got the *total* cause and the *total* effect. But, although this may well be so, I do not see that there is any logical necessity that it should be so.

We come now to the remaining postulate of the correlated variation of determinates.

Before stating this postulate it will be convenient to introduce a certain notation which will enable us to formulate it briefly and clearly. Let us suppose that $C_1 \ldots C_n$ is a total cause of which E is the total effect. Consider a certain one factor in this total cause, *e.g.*, C_r. I propose to denote the conjunction of the remaining factors $C_1 C_2 \ldots C_{r-1} C_{r+1} \ldots C_n$ by the single symbol Γ_{n-r}. The total cause can then be denoted by the symbol $C_r \Gamma_{n-r}$. Suppose now that a certain determinate value is assigned to each of the factors in Γ_{n-r}. We shall thus get a certain determinate value of Γ_{n-r}, and this may be denoted by γ^a_{n-r}. Let a certain determinate value of C_r be denoted by c^x_r. Then the determinate value of the total cause may be denoted by $c^x_r \gamma^a_{n-r}$. To this there will correspond a certain one determinate value of E. Let us denote this by $e^{x,a}_{r,n-r}$. We are now in a position to state our postulate.

(5) Let $C_1 \ldots C_n$ be a total cause of which E is the total effect. Select any one factor C_r from this, and assign to the remainder Γ_{n-r} any fixed value γ^a_{n-r}. Then, if there are *at least two* values of C_r, *e.g.*, c^x_r and c^y_r, which determine *different* values of E, *every* different value of C_r in combination with γ^a_{n-r} will determine a *different* value of E.

With the notation explained above the postulate can be stated very simply in the symbolism of *Principia Mathematica*. It will run as follows:

$$(\exists x, y).c^x_r \neq c^y_r.e^{x,a}_{r,n-r} \neq e^{y,a}_{r,n-r} : \supset_{a,r} : c^x_r \neq c^y_r$$
$$. \supset_{x,y} . e^{x,a}_{r,n-r} \neq e^{y,a}_{r,n-r}.$$

Now there are two other propositions which are logically equivalent to this postulate. The first is reached by taking the contra-positive of Postulate 5. We will call it (5a). It runs as follows:

(5a) Let $C_1 \ldots C_n$ be a total cause of which E is the total effect. Select any one factor C_r from this, and assign to the remainder Γ_{n-r} any fixed value γ_{n-r}^a. Then, if there are *at least two* values of C_r, e.g., c_r^x and c_r^y, which in combination with γ_{n-r}^a determine *the same* value of E, *every* value of C_r in combination with γ_{n-r}^a will determine *the same* value of E.

This can be put in the symbolism of *Principia Mathematica* as follows:

$$(\exists x, y).c_r^x \neq c_r^y . e_{r,n-r}^{x,a} = e_{r,n-r}^{y,a} : \supset_{a,r} : (x, y).e_{r,n-r}^{x,a} = e_{r,n-r}^{y,a}.$$

The second logically equivalent form of Postulate 5 may be called (5b). It is reached by substituting for the original hypothetical proposition the equivalent denial of a certain conjunctive proposition, in accordance with the general principle that 'if p then q' is equivalent to the denial of the conjunction 'p and not-q'. It runs as follows:

(5b) Let $C_1 \ldots C_n$ be a total cause, of which E is the total effect. Select any other factor C_r from this, and assign to the remainder Γ_{n-r} any fixed value γ_{n-r}^a. Then it cannot be the case *both* that there is a pair of values of C_r which in combination with γ_{n-r}^a determine *different* values of E, *and also* that there is a pair of values of C_r which in combination with γ_{n-r}^a determine *the same* value of E. This can be symbolised as follows:

$$\sim \{(\exists x, y).c_r^x \neq c_r^y . e_{r,n-r}^{x,a} \neq e_{r,n-r}^{y,a} : (\exists x, y)$$
$$.c_r^x \neq c_r^y . e_{r,n-r}^{x,a} = e_{r,n-r}^{y,a}\}.$$

I will now make some comments on this postulate. (a) It will be seen, on referring back to the first section of this paper, that (5) and (5a) are propositions of the form required to enable them to be used as major premises in demonstrative inductions. They are used as such by Mr. Johnson in his '*Figure of Difference*' and his '*Figure of Agreement*' respectively.

(b) It will be noticed that, when the conditions of (5) are fulfilled, not only is the *presence* of C_r relevant to the *presence* of E, but also the *variations* of C_r are relevant to the *variations* of E. Postulate 5 may therefore be called the 'Postulate of *Variational Relevance*'. When the postulate is put in the equivalent form (5a), and the conditions are fulfilled, the presence of C_r is relevant to the presence of E, but the variations of C_r are irrelevant to the variations of E. So, in this form, it may be called the 'Postulate of *Variational Irrelevance*'. An interesting example of variational irrelevance is furnished by Prof. H. B. Baker's discovery that gases which normally combine with explosive violence when a spark is passed through a mixture of them will not combine at all if they be completely dry. Thus the presence of *some* water is a necessary condition for any combination to take place in the assigned circumstances. But, granted that there is some water present, no difference in the amount of it seems to make any appreciable difference to the completeness or the violence of the combination which takes place when a spark is passed through the mixed gases.

(c) It must of course be clearly understood that, when the conditions of (5) are fulfilled, it follows only that variations of C_r are relevant so long as Γ_{n-r} is kept fixed *at the value* γ^a_{n-r}. For other values of Γ_{n-r} variations in C_r might be irrelevant. Similarly, when the conditions of (5a) are fulfilled, it follows only that variations in C_r are irrelevant so long as Γ_{n-r} is kept fixed *at the value* γ^a_{n-r}. For other values of Γ_{n-r} variations in C_r might be relevant.

(d) Finally we come to the question: 'Is this postulate true?' It seems to me quite certain that it is not. The fact is that Mr. Johnson, who first stated it, has altogether ignored the possibility of natural laws which take the form of periodic functions. Suppose there were a natural law of the form $E = C_1 \sin C_2$. Let C_1 be assigned a certain value. Take any value C_2^x of C_2. Then, for every value of C_2 that differs from this by an integral multiple of 2π, E will have the same value. On the other hand, for every value of C_2 which does not differ from this by an integral multiple of 2π, E will have a different value. Thus (5b) is directly contradicted. Nor is the kind of law which leads to these results at all fanciful. Such laws hold in electro-magnetism for alternating currents and the magnetic forces which depend on them. Thus the effect of the Postulate is to exclude all laws which take the form of periodic functions. And there is no *a priori* objection to such laws, whilst some important natural phenomena are in fact governed by laws of this kind.

It is worth while to remark that the existence of periodic laws answers in the affirmative a question which was raised and left unanswered in our comments on Postulate (4). The question was whether it is possible that a single determinate value of a total effect should correspond to an infinite plurality of alternative values of the total cause. In the case of periodic laws this possibility is realised. In our example, if C_1 be fixed, every one of the infinite class of values of C_2 which differ from each other by an integral multiple of 2π will determine one and the same value of E.

8. MR. JOHNSON'S 'FIGURES OF INDUCTION'

It only remains to explain and exemplify Mr. Johnson's 'Figures of Induction'. These are based on Postulate (3), *i.e.*, the Postulate of the Uniqueness of the Determinate Total Effect, and on one form or other of Postulate (5). The 'Figure of Difference' uses this postulate in its first form, *i.e.*, in the form of the Postulate of Variational Relevance. The 'Figure of Agreement' uses it in the second form (5a), *i.e.*, in the form of the Postulate of Variational Irrelevance. All the Figures also presuppose Postulate 4, *i.e.*, the Postulate of Variational Independence. And, since they all presuppose that a certain set of determinables has been shown to stand in the relation of total cause to a certain other set of determinables as total effect, they all presuppose the two postulates of Conjunctive Independence and of Smallest Sufficient Conditions. For these are involved in the arguments which are used in establishing laws of the Conjunction of Determinables. We will now consider the Figures in turn.

i. *Figure of difference*

The premises are as follows:

$C_1 \ldots C_n$ is a total cause of which E is the total effect. (a).

In a certain instance a certain determinate value $c_r^u \gamma_{n-r}^a$ is accompanied by a certain determinate value e of E. (b).

In a certain instance a certain determinate value $c_r^v \gamma_{n-r}^a$ is accompanied by a certain determinate value e' of E. (c).

c_r^u and c_r^v are different values of C_r; and e and e' are different values of E. (d).

The argument runs as follows:

From (a), (b), and Postulate (3) it follows that *every* instance of $c_r^u \gamma_{n-r}^a$ is also an instance of e.

From (a), (c), and Postulate (3) it follows that *every* instance of $c_r^v \gamma_{n-r}^a$ is also an instance of e'.

From these conclusions, together with (d) and Postulate (5), the following conclusion results: 'Corresponding to *each* value of $C_r \gamma_{n-r}^a$ there is a certain value of E, such that *every* instance of that value of $C_r \gamma_{n-r}^a$ is an instance of that value of E. And for *every* different value of $C_r \gamma_{n-r}^a$ the

THE PRINCIPLES OF DEMONSTRATIVE INDUCTION

corresponding value of E is different.' That is

$$c_r^x \neq c_r^y . \supset_{x,y} . e_{r,n-r}^{x,a} \neq e_{r,n-r}^{y,a}.$$

ii. *Figure of agreement*

The premises are as follows:
$C_1 \ldots C_n$ is a total cause of which E is the total effect. (a).
In a certain instance a certain determinate value $c_r^u \gamma_{n-r}^a$ is accompanied by a certain determinate value e of E. (b).
In a certain instance a certain determinate value $c_r^v \gamma_{n-r}^a$ is accompanied by the same determinate value e of E. (c).
c_r^u and c_r^v are different values of C_r. (d).
The argument runs as follows:
From (a), (b), and Postulate (3) it follows that *every* instance of $c_r^u \gamma_{n-r}^a$ is also an instance of e.
From (a), (c), and Postulate (3) it follows that *every* instance of $c_r^v \gamma_{n-r}^a$ is also an instance of e.
From these conclusions, together with (d) and Postulate (5a), the following conclusion results: 'Corresponding to *each* value of $C_r \gamma_{n-r}^a$ there is a certain value of E, such that *every* instance of that value of $C_r \gamma_{n-r}^a$ is an instance of that value of E. And for *every* value of $C_r \gamma_{n-r}^a$ the corresponding value of E is the same, viz., e.' That is

$$(x, y).e_{r,n-r}^{x,a} = e_{r,n-r}^{y,a} = e.$$

I will now make some comments on these two figures. The important point to notice is that each makes a *double* generalization by means of two different applications of demonstrative induction. The first generalises from a given *instance* of a given value to *all instances* of *that* value. This part of the argument rests on the Postulate of the Uniqueness of Determinate Total Effects. The second generalises from a given *pair of values* of a certain determinable cause-factor to *every pair of values* of that cause-factor. This part of the argument rests on the Postulate of Variational Relevance or Variational Irrelevance. The final result sums up both generalisations.

It may be remarked that, when we have the premises needed for the

Figure of Agreement, we can reach a more determinate conclusion than when we have the premises needed for the Figure of Difference. In the former case we know the determinate value of E which will be present in every instance in which any value of C_r is combined with γ_{n-r}^a. In the latter case we know only that a different determinate value of E will be present for each different determinate value of C_r combined with γ_{n-r}^a. We do not know what value of E will be correlated with each different value of $C_r\gamma_{n-r}^a$. To discover this we need to use the methods of *Functional Induction*; and this is a branch of *Problematic*, not of *Demonstrative*, Induction, and so falls outside the scope of this paper. Thus any complete inductive investigation begins and ends with Problematic Induction, and uses Demonstrative Induction only in its intermediate stages. It begins with Problematic Induction in order to establish Laws of the Conjunction of Determinables. In order to get these into the form of laws which express the relation of total cause to total effect it has to use the kind of deductive arguments which we considered in connexion with Necessary and Sufficient Conditions. In order to discover which factors in the total cause are variationally relevant and which are variationally irrelevant it has to use Mr. Johnson's figures, or something equivalent to them. And, in order to discover the detailed functional relation between variations in the total cause and variations in the total effect, it has finally to resort to a form of Problematic Induction.

iii. *Figure of composition*

The premises are as follows:

$C_1 \ldots C_n$ is a total cause of which E is the total effect. (a).

In a certain instance a certain determinate value $c_r^u \gamma_{n-r}^a$ is accompanied by a certain determinate value e of E. (b).

In a certain instance a certain determinate value $c_r^v \gamma_{n-r}^a$ is accompanied by a certain determinate value e' of E. (c).

In a certain instance a certain determinate value $c_r^w \gamma_{n-r}^b$ is accompanied by a certain determinate value e of E. (d).

The three values of C_r are all different, and e' is different from e. (e).

The argument runs as follows:

From (a), (b), and Postulate (3) it follows that *every* instance of $c_r^u \gamma_{n-r}^a$ is also an instance of e.

THE PRINCIPLES OF DEMONSTRATIVE INDUCTION

From (a), (c), and Postulate (3) it follows that *every* instance of $c_r^v \gamma_{n-r}^a$ is also an instance of e'.

From (a), (d), and Postulate (3) it follows that *every* instance of $c_r^w \gamma_{n-r}^b$ is also an instance of e.

Now it is impossible that γ_{n-r}^a should be the same as γ_{n-r}^b. For, if we first take (b) and (d) together, and then take either (b) and (c) together or (c) and (d) together, this would directly contradict Postulate (5b), in view of (e).

The final conclusion which results is this: 'Corresponding to *every* different value of C_r which, in conjunction with some value of Γ_{n-r}, determines the same value e of E there is a different value of Γ_{n-r}. And, in *every* instance in which a certain value of C_r is present along with e and some value of Γ_{n-r}, Γ_{n-r} will be present in the value that corresponds to this value of C_r.'

I must remark that Mr. Johnson's formulation of this figure at the bottom of page 225 of Part II of his *Logic* seems to me quite unsatisfactory. He there mentions only two instantial premises, whilst it is quite evident from the verbal statement of the figure which he makes earlier on the same page that a third instantial premise is essential to distinguish this from the Figure of Difference.

We can carry the argument a step further if we now add the premise (f) that γ_{n-r}^a and γ_{n-r}^b are known to differ *only* in the respect that a certain determinable C_s is present in the former in the value c_s^a and in the latter in the value c_s^b. In that case γ_{n-r}^a can be written as $c_s^a \gamma_{n-r-s}^\alpha$, and γ_{n-r}^b can be written as $c_s^b \gamma_{n-r-s}^\alpha$.

The final conclusion then runs as follows: 'Corresponding to *every* different value of C_r which, in conjunction with γ_{n-r-s}^α and with some value of C_s, determines the same value e of E there is a different value of C_s. And, in *every* instance in which a certain value of C_r is present along with e and some value of C_s, C_s will be present in the value which corresponds to this value of C_r.'

We may symbolise the value of E which is always present when $c_r^x c_s^a \gamma_{n-r-s}^\alpha$ is present by $e_{r,\,s,\,n-r-s}^{x,\,a,\,\alpha}$. The first clause of the above conclusion can then be symbolised as follows:

$$c_r^x \neq c_r^y . e_{r,\,s,\,n-r-s}^{x,\,a,\,\alpha} = e_{r,\,s,\,n-r-s}^{y,\,b,\,\alpha} = e : \supset_{x,\,y} . c_s^a \neq c_s^b.$$

The first clause of the conclusion which we reached before adding the

155

premise (f) may be symbolised as follows:

$$c_r^x \neq c_r^y . e_{r,n-r}^{x,a} = e_{r,n-r}^{y,b} = e : \supset_{x,y} . \gamma_{n-r}^a \neq \gamma_{n-r}^b.$$

iv. *Figure of resolution*

This figure is in a different position from the others. Here we have three observations which directly conflict with Postulate (5b) given the premise that $C_1 \ldots C_n$ is a total cause of which E is the total effect. It is evident that, in such a case, the only solution is to suppose that we were mistaken in believing that every factor in $C_1 \ldots C_n$ is simple. One at least of them must be a conjunction of at least two determinables, e.g., K_1 and K_2. The kind of premises which would lead to this conclusion are the following:

In a certain instance a certain determinate value $c_r^u \gamma_{n-r}^a$ is accompanied by a certain determinate value e of E. (b).

In a certain instance a certain determinate value $c_r^v \gamma_{n-r}^a$ is accompanied by a certain determinate value e' of E. (c).

In a certain instance a certain determinate value $c_r^w \gamma_{n-r}^a$ is accompanied by a certain determinate value e of E. (d).

c_r^u, c_r^v, and c_r^w are all different. And e' is different from e. (e).

It is evident that, if (b) and (d) be taken together, and if either (b) and (c) or (c) and (d) be taken together, there is a direct conflict with Postulate (5b). The only solution is to suppose that C_r is really a conjunction of two determinables, K_1 and K_2. In that case c_r^u may be $k_1 k_2$, c_r^v may be $k_1' k_2'$, and c_r^w may be $k_1'' k_2''$; and the contradiction will be avoided. It seems needless to pursue this into further detail.

Complete symbolic statement of the first three figures

I will bring this paper to an end by giving a complete symbolisation for the premises and the conclusions of the first three of Mr. Johnson's figures. Mr. Johnson's own symbolism seems to me to be very inadequate. For the present purpose we shall need two further bits of symbolism. (i) I will symbolise the premise that $C_1 \ldots C_n$ is a total cause of which E is the total effect by $C_1 \ldots C_n \rightarrow E$. (ii) We need a symbol for the statement that x is an instance of a conjunction of characteristics, A, B, C…Z. I shall denote this by [AB…Z] x. We are now in a position to deal with the figures of Difference, Agreement, and Composition.

THE PRINCIPLES OF DEMONSTRATIVE INDUCTION

Difference

$C_1 \ldots C_n \to E$ (a)

$[c_r^u \gamma_{n-r}^a e] p$ (b)

$\therefore [c_r^u \gamma_{n-r}^a] \xi \supset_\xi [e] \xi$ (by Postulate 3)

$[c_r^v \gamma_{n-r}^a e'] q$ (c)

$\therefore [c_r^v \gamma_{n-r}^a] \xi \supset_\xi [e'] \xi$ (by Postulate 3)

$c_r^u \neq c_r^v . e \neq e'$ (d)

$\therefore c_r^x \neq c_r^y . \supset_{x,y} . e_{r,n-r}^{x,a} \neq e_{r,n-r}^{y,a} : [c_r^x \gamma_{n-r}^a] \xi \supset_{\xi,x} [e_{r,n-r}^{x,a}] \xi$

(by Postulate 5).

Agreement

$C_1 \ldots C_n \to E$ (a)

$[c_r^u \gamma_{n-r}^a e] p$ (b)

$\therefore [c_r^u \gamma_{n-r}^a] \xi \supset_\xi [e] \xi$ (by Postulate 3)

$[c_r^v \gamma_{n-r}^a e] q$ (c)

$\therefore [c_r^v \gamma_{n-r}^a] \xi \supset_\xi [e] \xi$ (by Postulate 3)

$c_r^u \neq c_r^v$ (d)

$\therefore (x, y) . e_{r,n-r}^{x,a} = e_{r,n-r}^{y,a} = e : [c_r^x \gamma_{n-r}^a] \xi \supset_\xi [e] \xi$

(by Postulate 5a).

Composition

$C_1 \ldots C_n \to E$ (a)

$[c_r^u \gamma_{n-r}^a e] p$ (b)

$\therefore [c_r^u \gamma_{n-r}^a] \xi \supset_\xi [e] \xi$ (by Postulate 3)

$[c_r^v \gamma_{n-r}^a e'] q$ (c)

$\therefore [c_r^v \gamma_{n-r}^a] \xi \supset_\xi [e'] \xi$ (by Postulate 3)

$[c_r^w \gamma_{n-r}^b e] t$ (d)

$\therefore [c_r^w \gamma_{n-r}^b] \xi \supset_\xi [e] \xi$ (by Postulate 3)

$c_r^u \neq c_r^v . c_r^v \neq c_r^w . c_r^w \neq c_r^u . e \neq e'$ (e)

$\therefore c_r^x \neq c_r^y . e_{r,n-r}^{x,a} = e_{r,n-r}^{y,b} = e : \supset_{x,y} . \gamma_{n-r}^a \neq \gamma_{n-r}^b$

(by Postulate 5b)

Denote the value of Γ_{n-r} which corresponds to c_r^x by $_r^x \gamma_{n-r}^e$.
Then

$$[c_r^x \Gamma_{n-r} e] \xi \supset_\xi [_r^x \gamma_{n-r}^e] \xi.$$

Let
$$\Gamma_{n-r} = C_s \gamma^{\alpha}_{n-r-s} \quad (f).$$
Then
$$c^x_r \not\equiv c^y_r . e^{x,a,\alpha}_{r,s,n-r-s} = e^{y,b,\alpha}_{r,s,n-r-s} = e : \supset_{x,y} c^a_s \not\equiv c^b_s.$$

Denote the value of C_s which corresponds to c^x_r by $^x_r c^e_s$.
Then
$$^x_r \gamma^e_{n-r} = {^x_r} c^e_s \gamma^{\alpha}_{n-r-s}.$$
Whence
$$[c^x_r C_s \gamma^{\alpha}_{n-r-s} e] \xi \supset_{\xi,x} [^x_r c^e_s \gamma^{\alpha}_{n-r-s}] \xi$$
$$\supset_{\xi,x} [^x_r c^e_s] \xi.$$

Thus the complete final conclusion is:
$$c^x_r \not\equiv c^y_r . e^{x,a,\alpha}_{r,s,n-r-s} = e^{y,b,\alpha}_{r,s,n-r-s}$$
$$= e : \supset_{x,y} c^a_s \not\equiv c^b_s : . [c^x_r C_s \gamma^{\alpha}_{n-r-s} e] \xi \supset_{\xi,x} [^x_r c^e_s] \xi.$$

MECHANICAL AND TELEOLOGICAL CAUSATION

INTRODUCTION

Before embarking on a philosophical discussion of Causation it is desirable to draw certain distinctions. I begin by distinguishing between *Causal Propositions* and *Principles about Causation*. By a 'causal proposition' I mean any proposition which asserts of something that it is causally connected with something. Such propositions may be singular, *e.g.*, 'The death of Harold at Hastings caused the defeat of the English Army'; or they may be universal, *e.g.*, 'Friction causes rise of temperature'. The latter are called *Causal Laws*. By a 'principle about causation' I mean a general principle about causal propositions. Examples would be: 'Every event is causally determined', 'An effect and its cause must be manifestations of different determinate values of the same supreme determinables', and so on.

Next I will distinguish three questions. (1) Can causal propositions be *analysed*; and, if so, what is the right analysis of them? (2) Are there any causal propositions which I *know* or have grounds for *rationally believing?* (3) Do I know, or have I grounds for rationally believing, that there are some true causal propositions? In future I will use the phrase 'rationally cognize' for 'know or have grounds for rationally believing'.

It is very easy to confuse the second and the third questions with each other, but they are quite different. If the second is answered in the affirmative, it follows that the third must be answered affirmatively also. But the converse of this does not hold. I might rationally cognize the proposition that there are some true causal propositions, and yet there might not be a single causal proposition which I rationally cognize. Suppose, *e.g.*, that it were a self-evident principle about causation that every event is causally determined. Then I should know that there must be some true causal propositions. But there might still be no causal propositions which I rationally cognize. This example suggests that there is a fourth question to be added to our original three, *viz.*, (4) Are there any intuitively or demonstratively *a priori* principles about causation? It is clear that there might be such principles even if there were no intuitively or demonstratively *a priori* causal laws.

Let us now consider the connexion between the first question and the second. It might be that, if causal propositions were analysed in a certain way, it would necessarily follow that there would be no causal propositions which I rationally cognize. If I feel certain that this analysis is correct, and I see this consequence, I ought to admit that there are no causal propositions which I rationally cognize. If, on the other hand, I feel certain that there are some causal propositions which I rationally cognize and I see this consequence, I ought to reject this analysis, even though I cannot think of any alternative to it. But there is a third, and much more uncomfortable, possibility. I may feel quite certain that this is the right analysis so long as I do not notice that this anwser to the first question would compel me to give a negative answer to the second. I may feel equally certain that the second question must be answered in the affirmative so long as I do not notice that such an answer would compel me to reject what seems to be the right analysis. If I am in this situation, the only honest attitude for me to take is that of complete doubt and suspension of judgment about *both* questions. Similar remarks would apply, *mutatis mutandis*, to the connexion between the first question and the third or the fourth.

INDUCTION AND CAUSAL ENTAILMENT

I think that all the other symposiasts would claim to have rational cognition of some causal propositions, though I am not sure that they would all claim to know some causal propositions. *One* important line of argument, which is explicit in Prof. Stout's paper and is not questioned by Dr. Ewing or by Mr. Mace, may be fairly stated as follows. (i) In many cases past experience of a certain kind makes it rational for me to conjecture that a particular which I have not examined will have a certain characteristic ψ if it has a certain other characteristic ϕ. (ii) This would be impossible unless such past experience made it rational for me to conjecture that there is a causal law connecting the occurrence of ϕ in any particular with the occurrence in it of ψ. (iii) If the 'regularity analysis' of causal laws were correct, no past experience, however extensive or of whatever kind, would make it rational for me to conjecture that there is such a causal law. (iv) Therefore the regularity analysis of causal laws must be rejected. (v) There is one and only one alternative to the regularity analysis, *viz.*, what I will call the 'entailment analysis'. (vi) Therefore the entailment analysis of causal laws must be accepted. Professor Stout then draws certain consequences from this, which the other two symposiasts do not admit to follow. For the present I shall ignore this further step, which is peculiar to Prof. Stout. I will consider the rest of the argument, which is, so far as I know, accepted by the other two symposiasts.

One preliminary comment is obvious. Prof. Stout should have shown us quite clearly that, if the entailment analysis *is* accepted, past experience of the right kind and amount *will* make it reasonable for me to conjecture that there is a causal law connecting the occurrence of ϕ in any particular with the occurrence in it of ψ. Suppose that this cannot be shown. Or suppose it can be shown that, even if the entailment analysis be accepted, past experience *will not* make it rational for me to conjecture that there is such a causal law. Then, if the rest of the argument is valid, only two alternatives are open. Either: (a) I was mistaken in thinking that past experience will ever make it rational for me to expect that a particular which I have not yet examined will have ψ if it has ϕ; or (b) there must be some alternative to the regularity analysis beside the entailment analysis. This is not merely captious criticism. In the first place, McTaggart, who accepted the entailment analysis, professed to show that it does not provide a rational basis for inductive inference; and his argument is certainly not *prima facie* unsound. Secondly, W. E. Johnson, who rejected the regularity analysis, did not accept the entailment analysis. Having lodged this preliminary protest, I will turn to clauses (ii) and (iii) of the argument.

It seems to me quite certain that clause (ii), as it stands, is false. If past experience of a certain kind and amount will ever make it rational for me to conjecture that a *considerable percentage* of the as yet unexamined instances of ϕ are instances of ψ, and will make it reasonable for me to expect that the next one that I meet will be a *fair sample* of the class of instances of ϕ, this will suffice to make it reasonable for me to expect that the next instance of ϕ will be an instance of ψ. It is not in the least necessary that the past experience should make it rational for me to conjecture that there is a *causal law* connecting the occurrence of ϕ with that of

ψ. The artificial example of drawing counters from a bag, noting the colour, and conjecturing the colour of the next to be drawn, is enough to refute clause (ii). No one supposes that we have to assume that there is a causal law connecting the characteristic of being a counter in a certain bag with the characteristic of having a certain colour. Yet, if conjectures about the next instance can be justified in any case, they can most easily be justified in these artificial cases.

Nevertheless, I think that this step in the argument can be defended. Even in the artificial cases of bags and counters past experience justifies conjectures about the percentage of ϕ's which are ψ only on certain assumptions. We must assume that the ϕ's which were ψ will remain so, and that those which were not ψ will not become so. We must assume that the examined ϕ's were a fair sample of the whole contents of the bag; that the bag is not so large that there are some parts of it which we cannot reach; that the ϕ's which are ψ do not specially stick to our hands; and so on. Now, in the first place, some of these assumptions are justified only if we have already established certain laws of nature by induction. And, in the second place, some of them certainly break down when we try to extend the argument from the artificial case to the investigation of nature. It is certain that I cannot have observed any ϕ's which do not yet exist or have not yet happened: that I cannot have observed any which were very remote in past time or very distant in space; and so on. So defenders of clause (ii) might fairly say: 'Unless you can establish certain *causal laws* by induction you cannot justify some of the assumptions which you have to make in order to apply induction to artificial cases like bags of counters. And even then you cannot apply to nature the *statistical* inductive arguments which you can legitimately apply to the bags of counters; for the assumptions needed for such arguments clearly break down when applied to nature. Therefore, either inductive evidence will justify you in believing certain *causal laws of nature* or it will not justify you in believing *any* propositions either about nature or about artificial systems like bags of counters.' Let us henceforth take clause (ii) in this amended form.

We can now pass to clause (iii) in the argument, *viz.*, that, if the regularity analysis of causal laws were correct, no past experience, however extensive and however regular, would justify me in believing any suggested causal law. It is not necessary at the moment to go elaborately into the refinements with which the regularity analysis would have to be stated if it is to avoid certain *prima facie* objections to it. For our immediate purpose it will be enough to state it in the following rough outline. Any causal law is a statement of the form: '100 per cent of the instances of ϕ which have been, are, or will be, in the history of the universe, respectively have been, are, or will be, instances of ψ.' I shall shorten this into the more manageable form: '100 per cent of the ϕ's in nature are ψ.'

Now it is quite certain that, if my *only* premise is 'I have observed N ϕ's, and 100 per cent of them were ψ,' there is no valid form of argument by which I can either prove or render probable the conjecture that 100 per cent of the ϕ's in nature are ψ. But there are two very important points to be noticed here. (a) If this be granted, it follows *a fortiori* that there is no valid form of argument by which, from the same *single* premise, I could prove or render probable that the presence of ϕ in anything *entails* the presence in it of ψ. The proof of this is simply, and, so far as I can see, quite conclusive. It is as follows. The proposition 'The presence of ϕ in anything is inconsistent with the absence of ψ in it' entails, and is not entailed by, the proposition '100 per cent of the ϕ's in nature are ψ'. The former is therefore a logically more sweeping proposition than the latter. Now it is logically impossible that evidence which *will not* justify one in accepting as certain or probable the *less* sweeping of two propositions *will* justify one in accepting as true or probable the *more* sweeping of the two. If L is a true causal law, and the entailment view of causal laws is correct, it follows that the proposition which the regularity view offers as the analysis of L is *true*, though it is not the *analysis* of L. On the other hand, if the regularity view of causal laws is correct, it does not follow that the proposition which the entailment view offers as the analysis of L is true. Hence evidence which would not suffice by itself to prove or render probable the proposition which the regularity view takes to be the analysis of L must *a fortiori* be insufficient by itself to prove or render probable the proposition which the entailment view takes to be the analysis of L.

(b) Of course it remains possible that I may rationally cognize another premise P, such that the conjunction of P with the proposition 'I have observed N ϕ's, and 100 per cent of them were ψ' would justify me in conjecturing that the presence of ϕ in anything entails the presence of ψ in it. But then it also remains possible that I may rationally cognize another premise Q, such that the conjunction of Q with the proposition 'I have observed N ϕ's, and 100 per cent of them were ψ' would justify me *directly* in conjecturing that 100 per cent of the ϕ's in nature are ψ.

I propose henceforth to substitute for the phrase 'the presence of ϕ in anything entails the presence in it of ψ, the shorter phrase 'ϕ *conveys* ψ'. And I propose to substitute for '100 per cent of the ϕ's in nature are ψ' the shorter phrase 'ϕ *is always accompanied by* ψ'. This being understood, the discussion proceeds as follows.

Anyone who wants to use the argument which we are examining ought to substitute for clause (iii) the following set of four clauses. (iii, a) The empirical premise that 100 per cent of the N ϕ's which I have observed have been ψ does not, *by itself*, justify me in conjecturing that ϕ is always accompanied by ψ. (iii, b) *A fortiori* this empirical premise, *by itself*, does not justify me in conjecturing that ϕ conveys ψ. (iii, c) I *do not* rationally cognize any proposition Q, such that the conjunction of Q with my empirical premise *would* justify me in conjecturing that ϕ is always accompanied by ψ but *would not* justify me in conjecturing that ϕ conveys ψ. (iii, d) I *do* rationally cognize a certain proposition P, such that the conjunction of P with my empirical premise *would* justify me in conjecturing that ϕ conveys ψ and *would* therefore *a fortiori* justify me in conjecturing that ϕ is always accompanied by ψ.

My position, so far, is that I accept (iii, a) and insist that (iii, b) follows from it. The question for me turns, therefore on (iii, c) and (iii, d). In order to show the reader what sort of propositions I have in mind when I talk of P and Q, I will ask him to consider the two following propositions. (1) 'Every characteristic of any particular is *conveyed* by some other characteristic (simple or compound) of that particular.' (2) 'Every characteristic of any particular is *an invariable companion* of some other characteristic (simple or compound) of that particular.' Proposition (1) entails, but is not entailed by, Proposition (2). But this is quite compatible with Proposition (1) being *self-evident* to a certain person and Proposition (2) *not* being self-evident to him. Let us suppose that Proposition (1) is in fact self-evident to me, whilst Proposition (2) is not. If I found Proposition (1) self-evident, I should know that in each of the N instances of ψ which I have observed there must have been *some* characteristic or set of characteristics (not necessarily the same in all) which conveys ψ. Since all the N observed instances of ψ were also instances of ϕ, it might perhaps be legitimate to conjecture that it was ϕ which conveyed ψ in all the N observed instances. If so, it would follow that it is equally legitimate to conjecture that *any* instance of ϕ will be an instance of ψ. If Proposition (2) is *not* self-evident to me, I could not use it in a similar way to justify *directly* the conjecture that ϕ is invariably accompanied by ψ. But suppose that Proposition (2) *were* self-evident to me. Then I should know, as an immediate consequence of it, that in each of the N instances of ψ which I have observed there must have been *some* characteristic or set of characteristics (not necessarily the same in all) which is always accompanied by ψ. Since all the N observed instances of ψ where also instances of ϕ, it might perhaps be legitimate to conjecture that ϕ is a characteristic which is always accompanied by ψ. I should then have reached the same conclusion *directly*, instead of reaching it *indirectly* as a consequence of a previous conclusion about conveyance.

I am not of course, saying that either of these propositions (1) and (2) *is* self-evident, or that the first is and the second is not. Nor am I saying that the suggested arguments which use them as premises are valid. I am merely asking the reader to make certain suppositions on these points, in order that he may understand what I have in mind in clauses (iii, c) and (iii, d) of my amendment to clause (iii) of the argument which we are examining. It is now evident that anyone who uses this argument is bound to do two things. (a) He must indicate to us some general principle about the *conveyance* of characteristics in nature, which we can rationally cognize, and which, in conjunction with suitable empirical premises, will justify us in conjecturing that a certain characteristic (simple or compound) conveys a certain other characteristic. (b) He must show that we do not rationally cognize any general principle about the *invariable accompaniment* of characteristics in nature, which, in conjunction with suitable empirical premises, will justify us in conjecturing that a certain characteristic (simple or compound) is invariably accompanied in nature by a certain other characteristic. My position about this is that I agree as to (b), but am still anxiously awaiting enlightenment about (a).

I will now restate the argument in an amended form, and will indicate what I accept in it and what seems to me doubtful. (1) *No* fact of the form: 'I have observed N instances of ϕ and they have all been instances of ψ' will justify me in making a conjecture of the form 'A certain unobserved instance of ϕ is probably an instance of ψ', unless *some* facts of the first form will justify me in making conjectures of the form: 'Probably ϕ is always accompanied by ψ.' This clause has to be stated in the above rather complicated way in order to allow for the fact that one may be justified in strongly expecting that the next counter to be drawn from a bag will be red, in view of the observation that several have been drawn and have all been red, and yet may not be justified on the same evidence in conjecturing that all the counters in the bag are red. I accept clause (1), when thus stated.

(2) No fact of the form: 'I have observed N instances of ϕ and they have all been ψ' will *suffice by itself* to justify a conjecture of the form: 'Probably ϕ is always accompanied by ψ'. Some additional premise is needed; and, if I am to make use of it, I must rationally cognize it. I accept clause (2).

(3) I do not rationally cognize any proposition which, in conjunction with an empirical premise of this form, *would* justify me in making a conjecture of the form: 'Probably ϕ is always accompanied by ψ,' but *would not* justify me in making a conjecture of the form: 'Probably ϕ conveys ψ'. Speaking for myself, I admit clause (3).

(4) Empirical facts of the form under consideration *do* sometimes justify me in making conjectures of the form: 'A certain unobserved instance of ϕ is probably an instance of ψ'. I do not feel at all certain of this clause when I reflect on its implications, though I cannot help constantly acting as if I believed it.

(5) Therefore I *do* rationally cognize some proposition which, in conjunction with empirical facts of this form, *would* justify me in making conjectures of the form: 'Probably ϕ conveys ψ'. I do not feel at all certain of this conclusion. It is difficult enough to state any proposition which *would* do what is wanted of it if I *did* rationally cognize it. Something like Keynes's Principle of Limited Variety would seem to have the best credentials for the post. It is still more difficult to believe that one rationally cognizes any proposition which would do what is wanted of it.

Thus my position may be summed up as follows. I accept the first three clauses; and I admit that they, in conjunction with the fourth, *entail* the fifth. I therefore admit that the fourth *implies* the fifth. But the fifth seems so doubtful that the only result is to make me feel doubts, which I might not otherwise have felt, about the fourth. If someone should now say to me: 'After all, you may rationally cognize some proposition of the kind alleged in the conclusion of the argument, although you have never succeeded in disentangling it and getting it clearly stated,' I should of course, agree that this is possible. But the more heartily I agreed the more inclined I should be to go back on my acceptance of clause (3). My ground for admitting that I do not rationally cognize any principle which *would* justify me in conjecturing that ϕ is always accompanied by ψ but *would not* justify me in conjecturing, on the same empirical evidence, that ϕ conveys ψ, is simply that I *cannot think of* any principle which would answer these conditions and is rationally cognized by me. It is not that I can positively see that there *could not be* a rationally cognizable principle which fulfilled the positive and the negative condition. I think that Prof. Stout and Dr. Ewing would perhaps claim to see this. If so, the fact that they cannot formulate the principle which the conclusion asserts that they must rationally cognize would not cast any doubt on clause (3). It would therefore not tend to invalidate the argument for the conclusion that they do in fact rationally cognize such a principle. But suppose that one's *only* ground for accepting clause (3) is failure to formulate any principle answering to the conditions, and that there is no positive insight that the conditions *could not* be fulfilled. Then the admission that I might rationally cognize a principle without being able to disentangle it and formulate it weakens my ground for accepting clause (3) just as much as it strengthens the conclusion against an obvious *prima facie* objection.

This completes my discussion of the argument from the validity of inductive inferences to the entailment view of causation. It seems to me to be a very important argument, for the following reason. Suppose that all premises were accepted; and that the conclusion, which undoubtedly follows from them, were drawn. Then we should have done something much more important than merely showing that the regularity view of causal propositions is inadequate. Direct inspection and reflexion might, perhaps, convince many people that the regularity analysis fails to state what they have in mind when they are thinking of a causal law. But a person who admitted this might answer: 'Very well. But we have not the least *reason to believe* any causal law in any sense of that phrase but the sense which it would have if the regularity analysis were correct. Anything more, or anything different, that we may have in mind when we think of causal laws is, so far as one can tell, just baseless superstition which we must hand over to the genetic psychologist for explanation.' Now, if the argument which we have been discussing were accepted, this answer could not be made. The argument would show, not only that there is something involved in the notion of 'causal law' which the regularity view ignores, but also that we have *reason to believe* certain causal laws in a sense of 'causal law' which the regularity view fails to analyse.

REGULARITY ANALYSIS AND ENTAILMENT ANALYSIS

It seems to me fairly certain on inspection that I do not mean by 'causal laws' propositions of the form 'ϕ is always accompanied by ψ' limited by conditions about spatiotemporal and qualitative continuity and decked out with psychological frillings. And it seems to me fairly certain on inspection that I do mean by 'causal laws' propositions which involve the assertion that one proposition in some sense entails another, *i.e.*, that the truth of one is in some sense imcompatible with the falsity of the other. Whether I have any reason to believe that there are causal laws, and whether there are any causal laws which I have reason to believe, are, of course, two quite different questions from the question: 'What do I mean by a *causal law?*' And they are left quite open by the negative and the positive statements which I have just made in regard to the latter question. I propose at present to confine myself to some comments on the two alternative types of answer to this question about meaning or analysis.

(1) I have an uncomfortable feeling that the most impressive arguments *for* either kind of analysis are the objections *against* the other kind. The regularity analysis seems unplausible on inspection, and difficult to reconcile with the supposed validity of inductive arguments. So we are inclined to favour the entailment analysis until we look into it. When we do so we find perhaps that it does very little towards helping to justify inductive inference. And we certainly find that it is not very plausible to identify causal entailment with either of the two kinds of entailment which are commonly admitted to occur, *viz.*, the purely formal necessary connexion between the premise and the conclusion of a valid deductive argument, and the conveyance of extension by shape or of certain geometrical properties by certain others. Yet we not unnaturally hesitate to join W. E. Johnson in postulating a special kind of connexion between attributes, which is 'less necessary' than ordinary conveyance and 'less contingent' than constant accompaniment; to use phrases which are, perhaps, as absurd as they look, and yet do seem to express what one feels

to be needed. And so, to avoid the unplausibility of one form of the entailment view and the possible nonsense of the other form of it, we may be inclined to favour the regularity view. 'How happy could I be with neither,' as Macheath might have said if he had had to philosophize about causation.

(2) The other symposiasts have not explicitly distinguished between laws of coexistence of attributes in a substance and laws of sequence of events. It did not seem to be necessary to draw this distinction in discussing the connexion between induction and causal entailment, but it is desirable to draw it now.

Laws of co-existence, on the regularity view, are of the form: 'Every continuant in nature which has ϕ also has ψ'. On the entailment view they are of the form: 'The presence of ϕ in any continuant is incompatible with the absence of ψ from it'.

Laws of sequence on the regularity view, would be roughly of the following form. '(a) There have been, are now, or will be events which are manifestations of the characteristic ϕ in circumstances of the kind C. (b) Corresponding to any such manifestation of ϕ there has been, is now, or will be (according to whether it is past, present, or future) one and only one event which is a manifestation of a certain other characteristic ψ and which stands in a certain relation R to that manifestation of ϕ. (c) If e and e' be any two manifestations of ϕ in circumstances of the kind C, then the manifestation of ψ which corresponds to e, in accordance with the last clause, will be a different event from the manifestation of ψ which corresponds to e'.' This looks rather complicated, but I am sure that nothing less complicated will do. The relation R always involves immediate temporal sequence of the ψ-event on the ϕ-event with which it is correlated. In the case of purely physical causation it involves spatial coincidence or adjunction of the two events. In the case of purely mental causation it involves, perhaps, the concurrence of the two events in a single mind.

Laws of sequence, on the entailment view, would involve a proposition of the following form. '(a) If e is any manifestation of ϕ in circumstances of the kind C, it necessarily follows that there is one and only one event which is a manifestation of a certain other characteristic ψ and which stands in a certain relation R to e. (b) If e and e' be any two manifestations of ϕ in circumstances of the kind C, it necessarily follows that the manifestation of ψ which corresponds to e, in accordance with the last clause, will be a different event from the manifestation of ψ which corresponds to e'.' The same remarks apply to R as were made in the immediately previous paragraph.

(3) These statements of the two kinds of law, as they would be on the two types of analysis, bring out one point very clearly. The suggested analogy between admitted cases of non-formal entailment, in geometry, e.g., and the alleged cases of causal entailment breaks down almost completely for laws of sequence. All the admitted instances of non-formal entailment are instances of what I call 'conveyance' of one characteristic by another. They are all concerned with the *co-existence* of attributes in *substances*. Laws of sequence, on the entailment view, would all involve entailment between *instantial* propositions. For they would all assert, *inter alia*, that the *occurrence* of an event of one kind entails the *occurrence* of another event of a certain other kind. There is not the least analogy between such entailment and the conveyance of one characteristic of a substance by another.

(4) This leads me to think that, even if some form of the entailment view be true of laws of sequence, the form of it which is suggested by both Prof. Stout and Dr. Ewing is too stringent to be at all plausible. Dr. Ewing says: 'It would be possible in principle, with enough insight, to see what kind of effect must follow, from examination *of the cause alone* without having learnt by previous experience what are the effects of similar causes.' (My italics.) Prof. Stout makes similar assertions. Now, even if some form of the entailment view were true, this extreme consequence would not follow. Let us grant that a person who had observed a number of different manifestations of ϕ to be immediately followed by as many different manifestations of ψ, correlated each to each with the former, might be able to see that *any* manifestation of ϕ *must be* immediately followed by a correlated manifestation of ψ. It does not follow that, if only he were acute enough, he could see this *before* he had observed and reflected upon at least one instance of the sequence. Unless ψ and R are involved in the analysis of ϕ, as black is in that of negro, it seems obvious that he might have observed a manifestation of ϕ at a time when he could have had no idea of ψ or of R, and therefore at a time when he could not even have entertained the suggestion that manifestations of ϕ must be immediately followed by R-correlated manifestations of ψ. And yet, when experience had put him into a position to

understand and to contemplate this proposition, and had suggested it to him, he might be able to see that it must be true. In the case of the conveyance of one attribute of a substance by another attribute of it, the situation which I have envisaged could hardly arise. But, in the case of sequences, it obviously might arise.

(5) If we are to be fair to the regularity view, we must recognize that it could, and presumably would, distinguish between ultimate and derivative laws. Derivative laws are laws which follow as necessary consequences from one or more other laws. There are several different ways in which this could happen, and it will be worth while to enumerate some of the more important. (i) Suppose it is a law that ϕ is always accompanied in nature by ψ, and suppose it is a law that ψ is always accompanied in nature by χ. Then it necessarily follows that ϕ is always accompanied in nature by χ. This will be a derivative law as compared with the two which together entail it. (ii) Suppose it is a law that ϕ is always accompanied in nature by ψ. And suppose it is a necessary proposition that anything which had ψ would have ω. Then it necessarily follows that ϕ is always accompanied in nature by ω. This will be a derivative law as compared with the law which, in conjunction with the necessary proposition, entails it. The following are the two most obvious examples. (a) ψ might be a determinate under the determinable ω. Or (b) ψ might be a conjunctive characteristic of the form $\lambda\mu$; and it might be possible to see or to prove that anything which had *both* λ and μ would necessarily have ω, though not everything that had λ or everything that had μ would necessarily have ω. (iii) Suppose, as before, that there is a law that ϕ is always accompanied in nature by ψ. And suppose it is a necessary proposition that anything which had χ would have ϕ. Here again the two most obvious examples would be (a) if χ were a determinate under the determinable ϕ, or (b) if χ were a conjunction of two characteristics, λ and μ, about which we could see directly or prove that anything which had *both* of them would necessarily have ϕ. Here two different cases can arise, which it is important to distinguish.

(α) We might know that there are instances of χ in nature. Then we could at once infer the law that χ is always accompanied in nature be ψ. This law would be derivative, and would be of precisely the same kind as the other laws, ultimate or derivative, which we have so far considered. We may describe all the laws which we have so far considered as 'instantial laws'. By this I mean that they are not of the purely negative form: 'There are no particulars in nature which have ϕ and lack ψ': but they are of the form: 'There are particulars in nature which have ϕ, and none of them lack ψ'. If the regularity view be accepted, all *ultimate* laws of nature must be instantial, and many derivative laws will be instantial. Now it has been said that there are non-instantial laws of nature. E.g., it is a law that, if two perfectly elastic bodies were to collide, the total kinetic energy of the system would be the same before and after the impact; and this is not an instantial law, since there are no perfectly elastic bodies in nature. It has sometimes been made an objection to the regularity view that it leaves no room for non-instantial laws of nature. I will now pass to the second possible case, and I will show how far and in what way the regularity view can deal with non-instantial laws.

(β) We might not know that there are instances of χ in nature, or we might positively know that there are not. At the same time we may see that it would be impossible for anything to be an instance of χ without being an instance of ϕ. Under these circumstances we should be inclined to assert the law: 'If anything were χ it would be ψ'. What would this mean on the regularity view?

It could not be identified with the instantial universal proposition: 'There are instances of χ in nature and none of them lack ψ', for this is known to be false if it is known that there are no instances of χ in nature. It could not be identified with the purely negative proposition: 'There are no particulars in nature which have χ and lack ψ', for this is a mere truism if it is known that there are no particulars in nature which have χ. Lastly, it could not be identified with the proposition: 'The presence of χ in any particular would be incompatible with the absence of ψ from it', for this would commit us to the entailment view. What then does it mean on the regularity view?

The answer seems fairly plain. On the regularity view to say: 'If anything had χ it would have ψ' has the following meaning. It means that there is an instantial law of nature, or a set of such laws, such that it, or they, in conjunction with the *supposition* that there are instances of χ in nature, *formally entails* the proposition: 'There are instances of χ in nature and none of them lack ψ.' Take, *e.g.*, the proposition that kinetic energy would be unaltered in total amount by any collision between perfectly elastic bodies. What it means, on the regularity view, is this. There are instantial laws of nature (*viz.*, the Conservation of Momentum and the special laws about impact) which, when conjoined with the supposition that there are perfectly elastic bodies and that they sometimes collide, formally entail the instantial universal: 'There are cases of collision between perfectly elastic bodies, and in none of them is there any change in the total kinetic energy of the system.' The conclusion is false and one of the premises is false, but this is irrelevant. What we are concerned to assert is that this false conclusion is a necessary consequence of the conjunction of a certain false instantial supposition with certain true instantial laws of nature.

It is clear, then, that upholders of the regularity view can distinguish between ultimate and derivative causal laws, and that they can give a plausible interpretation to derivative non-instantial laws. On this point the differences between the upholders of the regularity view and the upholders of the entailment view are the two following. (i) On the regularity view the ultimate laws will be brute facts; whilst they will be intrinsically necessary propositions, true in all possible worlds, on the entailment view. (ii) On the regularity view all ultimate laws will be instantial; whilst, on the entailment view, there might be laws which were non-instantial and yet ultimate.

Plainly both parties could set before them the intellectual ideal of trying to find a minimal set of ultimate laws which would account for all the observed facts up to date, and they would both feel legitimate intellectual satisfaction in proportion as they reduced this minimal set more and more. If the regularity view is true, insight is being gained, in the negative sense that the number of independent brute facts to be accepted is being reduced, and in the positive sense that one is seeing necessary connexions between facts which are themselves contingent. On the entailment view, and on it only, a further kind of positive insight is conceivable, and it is therefore conceivable that an additional intellectual satisfaction could be enjoyed. For, on this view, the ultimate laws of nature would be intrinsically necessary propositions, 'holding in all possible worlds', and therefore it is conceivable that they might become self-evident to us, like the axioms of pure mathematics.

(6) So far we have confined our attention to laws of the crudest kind, *viz.*, those which assert merely of one *determinable* characteristic ϕ that it is invariably associated with a certain other *determinable* ψ. Suppose that

there are n determinates, $\phi_1, \phi_2, ..., \phi_n$, under the determinable ϕ, and that there are n determinates, $\psi_1, \psi_2, ..., \psi_n$, under the determinable ψ. Then the more refined kind of laws assert that every instance of any given determinate ϕ_r under ϕ is an instance of a certain one determinate under ψ. They assert that, if ϕ_r and ϕ_s are two different determinates under ϕ, then the determinate under ψ which invariably accompanies ϕ_r is different from the determinate under ψ which invariably accompanies ϕ_s. When the determinates under ϕ and under ψ are measurable magnitudes, laws reach a further degree of refinement. A law then asserts further that there is a certain one mathematical function, characteristic of ϕ and ψ, such that the number which measures any determinate ψ_r under ψ is this function of the number which measures the determinate ϕ_r which ψ_r invariably accompanies.

A law expressible by a mathematical equation of the form $\psi = F(\phi)$ would have to be stated as follows on the regularity view. '(i) Every determinate under ϕ is invariably accompanied by a certain one determinate under ψ. (ii) The determinate under ψ which invariably accompanies any *one* determinate ϕ_r under ϕ is different from the determinate under ψ which invariably accompanies any *other* determinate under ϕ. (iii) There is a certain one mathematical operation F, such that, if ϕ_r be any determinate under ϕ, and ψ_r be the determinate under ψ which invariably accompanies ϕ_r, then the number which measures ψ_r can be obtained by performing the operation F on the number which measures ϕ_r.' The same law, on the entailment view, would be stated by substituting throughout the word 'conveys' for the phrase 'is invariably accompanied by', and the phrase 'is conveyed by' for the phrase 'invariably accompanies.'

Now there are two remarks to be made about such laws. (a) It seems very doubtful whether there is any interpretation which can be put on clause (i) by the regularity view which would not make that clause either false or trivial. If it is interpreted instantly, it implies that there are in nature instances of *every* determinate under ϕ. Now in many cases the number of determinates under ϕ is enormous, and perhaps even infinite. It is extremely doubtful whether every possible pressure or temperature or volume has had or will have an instance in nature. Yet no one would think that this was any reason for doubting a well-established formula connecting the pressure, the volume, and the temperature of gases. If, on the other hand, it is interpreted non-instantially and yet in accord with the regularity view, it becomes trivial as regards any determinate under ϕ which has no instances in nature. If there are no particulars in nature which have the determinate ϕ_r, it follows of course that there are no particulars in nature which have ϕ_r, and lack a certain determinate ψ_r under ψ. But this is entirely trivial. We want to be able to say: 'If there *were* a particular which had ϕ in the form ϕ_r (which there is not), it *would* have ψ in the form ψ_r (though no particular in fact does)'. The entailment view can give a meaning to such statements which does not reduce them to trivialities. So far as I can see at present, the regularity view cannot.

(b) It is extremely difficult to suppose that we rationally cognize any principle which, in conjunction with suitable empirical premises, *would* justify us in believing a *functional regularity* but *would not* justify us in believing a corresponding *functional entailment*. I therefore agree with Prof. Stout that, unless there are laws of functional *entailment* and we have grounds for believing some of them on empirical evidence, we have no ground for believing any law of functional *regularity* on empirical evidence. On the other hand, even if there are laws of functional entailment, I could have no right to believe any of them on empirical evidence *alone*. I should have no ground to believe any of them unless I rationally cognize some principle which, in conjunction with suitable empirical premises, would justify such a belief. Unfortunately I cannot formulate any such principle which I could claim, with the least conviction, to be sufficient for this purpose and to be rationally cognized by me. My conclusion is as follows. Either (a) I do rationally cognize some principle which, in conjunction with suitable empirical premises, would justify me in believing certain laws of functional *entailment*, although I cannot elicit or formulate any such principle; or (b) no empirical evidence, however regular, varied, and extensive, gives me the slightest ground for believing a law even of functional *regularity*. I should tend, *prima facie*, to reject (b), as contrary to common-sense and to my own unquestioning convictions when not philosophizing about induction. But, when I realize that rejecting (b) entails accepting (a), I become more and more doubtful as to what I ought to hold.

ARE ANY CAUSAL PROPOSITIONS COGNIZED 'A PRIORI'?

We must begin by defining our terms. To say of a proposition p that it is 'known *a priori* by M' is equivalent to the following statement. '(a) The proposition *p* is a necessary one. (b) M knows that it is necessary; either by direct inspection of it, or by seeing that it is a necessary consequence of certain other propositions which he sees by direct inspection to be necessary.' Now the proposition *p* might be either *primary*, *i.e.*, not about any other proposition, or it might be *secondary*, *i.e.*, about some primary proposition *q*. If *p* is secondary, it might be of the form: 'The proposition *q* is more probable that not, given the datum *h*.' If I know this secondary proposition *a priori*, and I am acquainted with the datum *h*, I may be said to 'have *a priori* grounds for believing *q*'. We can now define the statement: 'M has *a priori* cognition of the proposition *x*.' It means: 'M either knows *x a priori* or has *a priori* grounds for believing *x*.'

We have seen that it is impossible that any *causal law* should be rationally cognized by induction unless certain *principles*, which we have not managed to formulate, are rationally cognized. I think it is evident that these principles, if rationally cognized at all, must be cognized *a priori*. But the question remains whether any *causal proposition* is an object of rational cognition to any human mind. The connexion between this question and the questions which we have been discussing in the earlier part of the paper is the following. If the entailment view of causal propositions is correct, they are propositions of such a kind that they might conceivably be known *a priori*, though it is, of course possible that none of them is in fact known *a priori*. For, on the entailment view, they are necessary propositions: and therefore it is conceivable that someone might be able to see or to prove the necessity of some of them. If the regularity view of causal propositions is correct, they are propositions of such a kind that they could not conceivably be known *a priori*. For, on that view, they are contingent propositions.

Having cleared up these preliminary matters, I will now make the following remarks on the question at issue.

(1) We are always liable to think that we have *a priori* knowledge of a synthetic proposition when really we are having such knowledge only of a trivial analytic proposition which we have confused with the former owing to some trick of language. The following would be an example. It might be said that I know *a priori* that it is impossible for me to be now remembering an experience *e* unless I formerly had the experience *e*. Now this is true in the sense that the phrase 'remembering *e*' is commonly used

in such a way that no experience of mine would be called a 'memory of e' unless I had formerly had the experience e. But, in that sense, it is analytic and trivial. If, on the other hand, it is taken to mean that I could not have had an experience which is psychologically indistinguishable from a memory of e unless I had experienced e beforehand, the proposition is synthetic and interesting, but is almost certainly not known *a priori*.

(2) Prof. Stout's claim to know *a priori* certain psychological causal propositions connected with inference seems to me very interesting. If he claimed to know *a priori* that the proposition: 'I am now believing p and seeing that it entails q' causally entails the proposition 'I shall believe q in the immediate future', several objections might be made. I might lose consciousness in the immediate future or be struck dead. Or the effect might be that I begin to doubt p or to doubt whether I saw that p entails q. Lastly, it seems to me conceivable that, if none of these things happened, I might still avoid believing q if it were very distasteful to me. But I think that the claim can be stated in a way which will avoid these objections. Suppose we put it as follows: 'So long as I am having the experience of believing p, of seeing that p entails q, and of considering whether q is true, it is impossible, that I should be having the experience of disbelieving q.' When the proposition is put in this form, it does seem to me to be self-evident; and, so far as I can see, it is not merely analytic. It is, of course, a proposition asserting *simultaneous* causation; whether an equally plausible example of an apparently self-evident and synthetic proposition involving causal *sequence* could be produced I do not know.

(3) Dr. Ewing holds that there are 'degrees of *a priori* intelligibility'. He is content to claim that we can see 'apart from experience' that some kinds of sequence (among mental events, at any rate) would be 'more intelligible than others'. Prof. Stout says that insight into necessary connexions among natural processes 'is in general, and perhaps always, only partial and imperfect'. But it is no more so, he adds, 'than, from the nature of the case, it ought to be in view of our inevitable ignorance of what actually takes place in causal process'.

It seems to me that this notion of 'degrees of *a priori* intelligibility' or of 'partial and imperfect insight into necessary connexions' needs more explanation than it has received. I am going to try to clear it up. Let us call any instance in which a number of conditions $c_1 \, c_2 \ldots c_n$ are simultaneously fulfilled in a certain relation S to each other a 'concurrence' of these conditions. Let us suppose that any concurrence of $c_1 \, c_2 \ldots c_n$ entails the simultaneous or immediately subsequent occurrence of one and only one event of a certain kind e which stands in a certain relation R to that occurrence. Let us further suppose that, if C and C' are two different concurrences of these conditions, the R-correlated e-event whose occurrence is entailed by C is different from the R-correlated e-event whose occurrence is entailed by C'. Finally, let us suppose that there is no selection from $c_1 \, c_2 \ldots c_n$, such that the propositions enumerated above are true of this selection. Then I shall say that $c_1 \, c_2 \ldots c_n$ are a 'Smallest Sufficient Condition' of e. I shall say that any one of these conditions, or any selection consisting of several of them, is a 'Relatively Necessary Condition' of e.

Now it is possible that there might be a number of alternative Smallest Sufficient Conditions of e. If there is any condition common to all the Smallest Sufficient Conditions of e, I shall call it an 'Absolutely Necessary Condition' of e. It is quite possible that there should be no Absolutely Necessary Condition of e. On the other hand, it is possible that there might be several Absolutely Necessary Conditions of e. If so, I shall call the most numerous set of conditions common to all the Smallest Sufficient Conditions of e 'The Greatest Absolutely Necessary Condition' of e. If e should have only one Smallest Sufficient Condition, every factor in this will be an Absolutely Necessary Condition of e, and e's only Smallest Sufficient Condition will be identical with e's Greatest Absolutely Necessary Condition.

Now even if e has *several* alternative Smallest Sufficient Conditions, and if it has *no* Absolutely Necessary Condition, it might still be possible to arrange e's Relatively Necessary Conditions in what I will call an 'order of dispensability.' Suppose that e has N alternative Smallest Sufficient Conditions, and has no Absolutely Necessary Condition. Suppose that a certain condition c_1 is present in m of these, and that a certain other condition c_2 is present only in k of them, where k is less than m. Then there would be a perfectly good sense in saying that whilst c_1 and c_2 are *both* relatively, and *neither* absolutely, necessary conditions of e, c_2 is a 'more dispensable' condition of e than c_1 is. To be an *absolutely* necessary condition of e is the same as to be a condition of e which has *zero* dispensability.

Now what is called 'The Law of Universal Causation' may be stated as follows. '(i) Every *kind* of event has one or more Smallest Sufficient Conditions. (ii) Every *occurrence* of an event of a given kind e is due to one and only one occurrence of one and only one of the Smallest Sufficient Conditions of e. (iii) If E and E' are two *different* occurrences of an event of a given kind e, then the occurrence to which E is due and the occurrence to which E' is due are *different* occurrences of the same or of different Smallest Sufficient Conditions of e.' This is a principle about causation which many people would claim to find self-evident, even if they did not claim to find any causal proposition self-evident. I have defined 'Smallest Sufficient Condition', and the other notions which are correlated with it, in terms of the entailment view of causation. But it would be possible, no doubt, to define them in terms of the regularity view; and a person who accepted the regularity view might find the Law of Universal Causation self-evident just as much as a person who accepted the entailment view. On the other hand, a person who accepted the entailment view, and who further claimed to find certain causal proposition self-evident, might nevertheless not find the Law of Universal Causation self-evident.

We must now apply these considerations to the question of 'degrees of insight' into causal connexions. Let us assume that a person accepts the entailment view of causation and finds the Principle of Universal Causation, when stated in terms of entailment, self-evident. The former assumption is certainly true of all the other symposiasts; they have not made any explicit statement on the second point, but I think it is safe to assume that Prof. Stout and Dr. Ewing, if not Mr. Mace, do find the Principle of Universal Causation self-evident when stated in terms of entailment.

Consider the following five propositions. (i) 'c is a relatively necessary condition of e.' This is equivalent to the proposition: 'There is *at least one* Smallest Sufficient Condition of e which contains c as a factor.' (ii) 'c is an absolutely necessary condition of e.' This is equivalent to the proposition: '*Every* Smallest Sufficient Condition of e contains c as a factor.' (iii) '$c_1 c_2 \ldots c_m$ is the Greatest Absolutely Necessary Condition of e.' This is equivalent to the proposition: '$c_1 c_2 \ldots c_m$ are all contained in every Smallest Sufficient Condition of e, and *no other* factor is contained in every Smallest Sufficient Condition of e.' (iv) '$c_1 c_2 \ldots c_n$ is a Smallest Sufficient Condition of e.' This is equivalent to the proposition: 'Each of the conditions $c_1, c_2, \ldots c_n$ is a relatively necessary condition of e, and their concurrence in a certain relation R to each other is a sufficient condition of e.'

(v) '$c_1 c_2 \ldots c_n$ is the only Smallest Sufficient Condition of e.' This is equivalent to the proposition: 'Each of the conditions $c_1, c_2 \ldots c_n$ is an absolutely necessary condition of e, and their concurrence in a certain relation R to each other is a sufficient condition of e.'

It is evident that (i) is a weaker proposition than (ii) and that (ii) is weaker than (iii). It is also evident that (iv) is weaker than (v) and that (iii) is weaker than (v). I do not think that any direct comparison can be made between (iv) and either (ii) or (iii). But it follows immediately that (i) is the weakest and (v) is the strongest of all these propositions.

The mildest possible claim to *a priori* knowledge of causal laws would be the claim to know *a priori* some propositions of the form (i). This claim is certainly made by Prof. Stout and Dr. Ewing. The boldest claim would be the claim to know *a priori* some propositions of the form (v). I do not think that any of the other symposiasts make this claim. An important intermediate claim would be to know *a priori* some propositions of the form (iv). This would involve knowing *a priori* some propositions of the form (i). For, if I know *a priori* that $c_1 c_2 \ldots c_n$ is a Smallest Sufficient Condition of e, I *ipso facto* know, with regard to each of the factors, that it is a relatively necessary condition of e. I am not sure whether Dr. Ewing claims to know *a priori* any proposition of the form (iv). But the example of the psychological law about inference would seem to imply that Prof. Stout claims to know some propositions of this form *a priori*.

Now the growth of insight into causal connexions, which Prof. Stout talks about, might take two forms. It might be *extensive*, *i.e.*, we might get to know *a priori* more propositions of a certain form than we knew before. Or it might be *intensive*, *i.e.*, we might pass from knowing *a priori* only propositions of a weaker form to knowing *a priori* certain propositions of a stronger form. Even if we do not do this, we might pass to knowing *a priori* that certain propositions of a stronger form are more and more highly probable. I am not sure whether Prof. Stout would claim that insight grows intensively as well as extensively. And, if he claims that it grows intensively, I am not sure whether he would claim that this intensive growth of insight is of the first kind or that it is only of the second.

(4) Propositions of the form: 'c is a relatively necessary condition of e' might be rationally cognized in two different ways, which might be called 'the *direct* way' and 'the *indirect* way' respectively. And the direct way

might take two different forms. (i) It might be that, without ever having observed a transaction in which an occurrence of e was due to a Smallest Sufficient Condition in which c was a factor, I could know *a priori* that c is a relatively necessary condition of e. If so, I should know this in the direct way. (a) Supposing this to be possible, I might be able to know *a priori* that c is a relatively necessary condition of e merely by reflecting on c and without needing to have observed an occurrence of e. Dr. Ewing seems to claim such knowledge. (b) It might be that I should need to have observed occurrences of e in order to give me the idea of e, but that, when I reflect on *both* c and e, I can know *a priori* that c is a relatively necessary condition of e. This seems to me to be a much more reasonable claim. These are the two forms of the direct way of rationally cognizing that c is a relatively necessary condition of e.

(ii) Suppose I observe a certain occurrence of e, and suppose I know the Law of Universal Causation. Then I know that this occurrence of e must have been simultaneous with or immediately successive to a certain one occurrence of one or other of e's Smallest Sufficient Conditions, and that this occurrence of e must stand in a certain relation R to this occurrence of this Smallest Sufficient Condition of e. Suppose I know that the relation R involves spatial adjunction or coincidence between the occurrence of an event and the occurrence of its Smallest Sufficient Condition in the case of physical events, and that it involves occurrence in the same mind in the case of mental events. If e is an event of a physical kind, suppose that a certain condition c was fulfilled in the immediate neighbourhood of this occurrence of e just before it happened. And suppose that, so far as I know, the situation in the immediate neighbourhood of this occurrence of e immediately before it happened differed from the situation in this neighbourhood a little while ago *only* by the fulfilment of c. Then it would be reasonable to conjecture that c is a relatively necessary condition of e. This could never be more than a reasonable conjecture. For, even if I *know* that only circumstances in the immediate neighbourhood of a physical event can be conditions of its occurrence, I could never *know* that the observed fulfilment of c immediately before the occurrence of e was the *only* change that had taken place in the immediate neighbourhood of this occurrence of e immediately before e happened.

This is what I mean by the 'indirect way' of getting rational cognition of a proposition of the form: 'c is a relatively necessary condition of e'.

It certainly cannot be said to give *a priori* knowledge of any causal proposition. But it can be said that it gives us rational belief in certain causal propositions, which *is* based partly on *a priori* knowledge of causal principles and *is not* reached by problematic induction.

CAUSATION AND CONATION

I propose to say very little on this topic, partly because I have had to say so much about the other parts of our subject, and partly because I find myself in almost complete agreement with the remarks which Dr. Ewing and Mr. Mace have made about Prof. Stout's theory. What I wish to add is this.

In defining the notion of 'Smallest Sufficient Condition', and the other notions connected with it, we had to mention two kinds of relation, S and R. S is the relation which a number of simultaneously fulfilled conditions must have to each other if they are to be factors in a single occurrence of a single Smallest Sufficient Condition. It might be called a 'Co-operative Bond'. R is the relation in which each occurrence of e stands to one and only one occurrence of some one of e's Smallest Sufficient Conditions; each different occurrence of e being correlated by R with a different occurrence of some one or other of e's Smallest Sufficient Conditions. It might be called a 'Consecutive Bond'. It is plain that the Co-operative Bond involves more than mere spatial coincidence or adjunction in the case of purely physical conditions, and that it involves something more than mere occurrence in one and the same mind in the case of purely mental conditions. Similar remarks apply to the Consecutive Bond.

Now any ideas that we may have of co-operative bonds and consecutive bonds must presumably be derived from instances in which we were acquainted with an occurrence of a Smallest Sufficient Condition followed immediately by the occurrence of the event which it caused. Plainly one's own conative processes are the most striking and important instances of this kind of process with which we are acquainted. It is therefore highly plausible to hold that we derive from our acquaintance with them the idea of a number of factors co-operating or conflicting with each other and thus forming a single Smallest Sufficient Condition. And it is plausible to hold that we derive from the same source the idea of a certain event being *the* event which is determined by a certain occurrence of a certain set of co-operating and conflicting simultaneous conditions.

It is from the qualities and relations and changes of particulars with which we are acquainted that we must ultimately derive all the ideas by which we think of things and processes with which we are not acquainted. The ideas of spatial characteristics, of extensible qualities, and of motion and qualitative change, which we ascribe to physical objects, are derived ultimately from the spatial characteristics, the sensible qualities, and the sensible motion and qualitative change, which are manifested to us by the sensa that we sense. Similarly, when we think of physical things and processes as causal factors which co-operate and conflict and thus constitute total causes of certain physical events, the ideas which we use must ultimately be derived from instances of co-operation, conflict, and consequence with which we are acquainted. And the most striking, if not the only, instances with which we are acquainted are our own conative processes, whether successful or thwarted. Therefore, when we think of the external world under dynamical categories, we are no doubt using conceptions derived ultimately from our acquaintance with our own conative processes; just as, when we think of the external world under spatial and kinematic and qualitative categories, we are using conceptions derived ultimately from our acquaintance with visual, tactual, auditory, and other sensa. Both procedures are psychologically inevitable; and, if *either* is epistemologically justifiable, there is no reason to think that the one is so and the other is not.

On the other hand, we must remember how extremely remote a concept, which is ultimately derived from certain features in objects with which we are acquainted, may be from the sensible or introspectable characteristic which is its ultimate source. Contrast, *e.g.*, the notion of a generalized non-Euclidean N-dimensional manifold with the visual field and the spatial characteristics which it presents to our inspection. The co-operative and consecutive bonds and their terms, which are involved in purely physical causation, may be as remote from those which we find in our own conative processes as the generalized non-Euclidean N-dimensional manifolds of the abstract geometer are from the visual fields which are the ultimate source of his spatial concepts.

CRITICAL NOTICE:

Wahrscheinlichkeit, Statistik, und Wahrheit. By R. VON MISES. Second Edition. Wien: Julius Springer, 1936. Pp. VIII + 282.

The first edition of this work was published in 1928. It now re-appears, in a considerably enlarged form, as Vol. III of the series *Schriften zur wissenschaftlichen Weltauffassung*, edited by Prof. Frank of Prague and the late Prof. Schlick of Vienna. The author is a very distinguished mathematician, formerly of Berlin and now professor at Istanbul. So far as I can discover, the first edition was not noticed in *Mind*. For this reason and because the book contains an extremely clear and able statement of one form of the Frequency Theory of Probability by an acknowledged expert in the technique of the subject, I propose to review it in some detail.

The book consists of six divisions. Each is called a 'lecture'; but they must have been considerably expanded from their original length. The first four of them contain the statement and explanation and defence of von Mises' theory; the other two are accounts of the application of the calculus of probability to statistics and the errors of observations and to physics. The first lecture deals with the definition of 'probability', and the second with the elements of the calculus of probability. In the third von Mises considers critically alternative views to his own, and tries to deal with the arguments of opponents and the alleged improvements suggested by half-converted friends. Plainly there is some close connexion between a frequency-theory of probability and those theorems which may be grouped together under the head of 'the laws of great numbers'. In the fourth lecture von Mises considers carefully the meaning of these theorems and their precise relation to his frequency-definition of 'probability'.

The essential points in Lecture I are the following: (i) The word 'probability' may be compared, *e.g.*, with the word 'work', in so far as it is undoubtedly used in many different senses in ordinary life, and it is hopeless to look for a definition of either which will both cover all these senses and mark out something capable of measurement and mathematical treatment. The proper course is to begin by attending to those regions in which the word 'probability' is admittedly used in a sense in which it can be and has been made the subject of a calculus. These are games of chance, insurance, and certain mechanical and physical problems. This clear central nucleus is surrounded by a penumbra of borderline cases, such as the credibility of witnesses. In the outer darkness, and explicitly excluded by von Mises from consideration, come such usages as the 'probability' of an historical narrative, the 'inner probability' of a work of art, and so on. I suppose that he would therefore exclude from consideration the 'probability' of an alleged law of nature, such as the conservation of energy, and the 'probability' of a scientific theory, such as Einstein's theory of gravitation or the nebular hypothesis.

(ii) We must next try to discover what is common and peculiar to the cases that fall within the clear central nucleus. According to von Mises we find two such characteristics, one fairly obvious and the other less so. (a) We have a certain clearly delimited class of observable phenomena, *e.g.*, throws with a die, which are very numerous and can be conceived to become indefinitely more numerous as time goes on. Each member of this class must manifest *some one*, and cannot manifest *more than one*, of a certain set of alternative characteristics. *E.g.*, in the case of the fall of a die, the characteristics are a 1 uppermost or a 2 uppermost or ... a 6 uppermost. The relative frequencies with which these various alternatives have been manifested can be determined at any moment, and it is conceived that each of them would approach indefinitely nearly to a certain limiting value as the total number of observed members of the class was indefinitely increased. (b) The frequencies with which the various alternatives are manifested among the members of a class might approach to limiting values in the way described above, and yet the following situation might exist. There might be one or more general rules for choosing infinite sub-classes out of the original infinite class, such that in these selected sub-classes the limiting frequency of a given alternative would be *different from* its limiting frequency in the original class. Now the second condition is that this possibility must be ruled out. The original class must be such that the limiting frequency for any alternative is *the same* for the class as a whole and for any infinite selection from it, provided only that the question whether a given individual does or does not fall into the selected subclass is independent of the particular alternative which *it* manifests. Von Mises refers to this second condition as the 'principle of the Impossibility of a Gambling System' or the 'principle of Insensitivity to Place Selection' (*Stellungauswahl*). We will call it 'Randomness'. He defines a 'Collective' (*Kollektiv*) as any class which answer to these two conditions.

(iii) 'Probability', as used by von Mises, has a meaning only in reference to collectives. The minimum intelligible statement predicating a probability is of the form 'The probability of the occurrence of alternative *a* in the collective C_A is *p*'. And this means what is meant by the statement 'C_A is a collective every member of which is a manifestation of some one of the alternatives $a, a', a'', \ldots$; and the limiting value of the ratio of the number of its members which manifest *a* to the total number of its members is *p*'.

If we consider a certain particular die, *e.g.*, the relevant collective will be the past, the present, and all the possible future throws with *that* die. With von Mises' definition it would be sensible to ask 'What is the probability of throwing a 6 with *that* die?' But, so far as I can see, it would be meaningless to ask 'What is the probability that I, who am now just about to throw that die, shall throw a 6 with it on this occasion?' The case of vital statistics would seem to be somewhat different, since each man can die but once. Here the collective might be, *e.g.*, Englishmen reaching the age of 40 during 1937, considered in respect of the two alternatives of surviving or not surviving their 41st birthday. It seems to me that the notion of a collective, answering to von Mises' two conditions which both involve infinity and limits, can hardly be regarded as a legitimate extrapolation from the observable class in this case, even if it can in the case of throwing a die.

Be this as it may it would be meaningless, on von Mises' definition, to ask 'What is the probability of *Mr. Smith*, who became 40 in 1937, surviving to *his* 41st birthday?' This is admitted and asserted by von Mises, but he uses an argument which is really relevant to a different point. The argument is that Mr. Smith, beside being an English *man*, is an English *human being*, is a *European* human being, and so on. Now the statistics for persons of 40 surviving to their 41st birthday are different for all these different classes, and Mr. Smith is equally a member of all of them. Why single out the statistics for one of them, *viz.*, the class of

VON MISES 'WAHRSCHEINLICHKEIT, STATISTIK, UND WAHRHEIT'

English men, rather than the statistics for another of them, as 'the probability that Mr. Smith will survive to his 41st birthday'? If you answer that it is unreasonable to take the statistics of a less determinately delimited class when you can get those of a more determinately delimited class, why stop at the class of English men? Mr. Smith may be a Yorkshireman, an Etonian, and a Plymouth Brother, besides being an English man. If you go far in this direction, you will define a class of which he is the only known member, and then the notion of limiting frequency will be completely inapplicable. My comment on this argument is twofold. In the first place, it is not needed in order to show that it is meaningless to talk of the probability of a particular event on von Mises' definition of 'probability'. This is immediately obvious from the definition. Secondly, if a person does attach a meaning to 'probability' as applied to particular events, all that the argument will teach him is what he knew already, *viz.*, that the must never talk of *the* probability of an event without qualification, but must always talk of its probability with respect to such and such data. There is nothing in the argument to prevent such a person from saying that the probability of Mr. Smith surviving to his 41st birthday, relative to the datum that he is an Englishman of 40 and to that alone, is measured by the frequency with which Englishmen of 40 have been found to survive to 41.

(iv) The fact that there are collectives, in von Mises' sense, is an empirical fact. The evidence for the existence of limiting frequencies in games with dice, cards, etc., is provided by the experience of gamblers, proprietors of casinos, governments holding lotteries, and so on. The evidence that these limiting frequencies are the same for all selections which fulfil von Mises' conditions is provided by the failure of all gambling 'systems'. On p. 16 von Mises says that the probability of a certain die throwing a 6, as defined by him, is 'a physical property of the die, of the same kind as its weight, its thermal conductivity, etc.' I think it is plain that these assertions are highly questionable; but I shall defer consideration of them until we have seen what von Mises has to say about the laws of Great Numbers, which are likely to be relevant in this connexion.

We can now pass to the second Lecture, which is concerned with the objects and methods of the Calculus of Probability. The general problem of the calculus may be stated as follows: 'You are given the probabilities for the various alternatives in certain collectives. You are asked to infer the probabilities for the various alternatives in certain other collectives *derived from* the former.' It is no part of the business of the calculus to provide the original probabilities; these must be supplied by observation or postulated hypothetically. To think otherwise is to make a mistake about the calculus of the kind which a person would make who confused geometry with mensuration. In every probability-calculation both the premises and the conclusion are statements of probabilities. Lastly, we must remember that the probabilities 0 and 1, on von Mises' theory, do not mean 'certainly not' and 'certainly', respectively. They mean only that the frequency of a certain alternative in a certain collective tends to 0 or to 1, respectively, as the number of terms is indefinitely increased.

The question that remains is 'What is meant by *deriving* a collective from other collectives, and how is it done?' Von Mises says that the process of derivation has four and only four fundamental forms, and that any particular case can be reduced to a single or a repeated application of one or more of these four procedures. He calls them *Selection, Mixing, Partition* and *Combination*. I will now explain what he means by them.

(i) *Selection*. This consists in selecting an infinite class, in accordance with some rule, from the members of a collective, and considering the probability of the *same* alternatives within the selected class. It follows from the definition of a collective that the probabilities are unchanged.

(ii) *Mixing*. Here we consider the same set of terms as before, but we take as a single alternative a disjunction of several of the original alternatives. Thus the original collective might be the throws of a certain die, considered in respect of the six alternatives 1, 2, 3, 4, 5, or 6 uppermost, and the probabilities of these might by $p_1, p_2, ..., p_6$, respectively. The derived collective might be the throws with the same die, considered in respect of the two alternatives odd or even uppermost. The first of these is a disjunction of the original alternatives 1, 3, and 5; and the second is a disjunction of the original alternatives 2, 4, and 6.

The rule for calculating the new probabilities in such cases is, of course, the Addition Rule. Von Mises remarks that this rule is often carelessly formulated. It is often said that the probability of (p or q) is equal to the sum of the probabilities of p and of q, provided that p and q are mutually exclusive. He points out that the probability of dying in one's 40th year or getting married in one's 41st year is not the sum of the probabilities of dying in one's 40th year and getting married in one's 41st year, although the alternatives are mutually exclusive. The condition which must be added is that one and the same *collective* is under consideration throughout. (On the Keynes–Johnson theory the corresponding condition would be that one and the same *datum*, e.g., the proposition h, must be considered throughout.)

(iii) *Partition*. The essential point of this may, I think, be put most clearly as follows. Suppose that, in your original collective, a certain set of n mutually exclusive and collectively exhaustive alternatives, $a_1, a_2, ..., a_n$ was considered. Form a new collective by excluding from consideration every member of the original collective which manifests any of the alternatives $a_{m+1} ... a_n$, and consider this in respect of the limiting frequencies of the remaining alternatives $a_1, a_2, ..., a_m$. The rule here is that, if p_1, $p_2, ..., p_m$ be the original probabilities for the alternatives $a_1, a_2, ..., a_m$, respectively, then the new probability for each will be got by dividing its old probability by the sum $p_1 + p_2 + ... + p_m$.

It seems to me rather futile to offer this as a fundamental procedure in the calculus. It is easy to show that the rule is a consequence of applying the general principle of inverse-probability to a certain very simple special case. And the rule of inverse-probability is itself an immediate consequence of the rule of multiplication which von Mises introduces later in connexion with what he calls 'Combination'. The proof of these statements is as follows. The multiplicative rule, stated in the Keynes–Johnson notation, is $(x.y)|h = (x|h)(y|xh) = (y|h)(x|yh)$. From this there immediately follows the rule of inverse-probability, viz., $x|yh = (x|h) \times (y|xh) \div (y|h)$. In order to get von Mises' rule of 'Partition' we have merely to substitute the disjunction $a_1 \vee a_2 \vee ... \vee a_m$ for y and to substitute a_1, e.g., for x. Then $x|h = p_1$; $y|xh$ obviously $= 1$, since it is the probability of an alternative proposition given that one of the alternants is true; and $y|h = p_1 + p_2 + ... + p_m$, since the alternatives are by hypothesis mutually exclusive and are being considered with respect to the same datum. So von Mises' rule of Partition follows at once.

(iv) *Combination*. Suppose we start with two collectives, C_A and C_B, for which the alternative possibilities are respectively $a_1, a_2, ..., a_n$ and $b_1, b_2, ..., b_m$. Let R be any relation which correlates the terms of C_A and C_B with each other in pairs. Consider the class each member of which is a *pair* of correlated terms, one from C_A and the other from C_B. Since the C_A-constituent of any such pair has the n alternatives $a_1 ... a_n$ open to it, whilst the C_B-constituent of the same pair has the m alternatives $b_1 \ldots b_m$ open to it, and each of the former could be combined with each of the latter, the terms of our new class can be considered in respect of the $n.m$ conjunctive alternatives of the form $a_1 b_1, a_2 b_1 \ldots a_n b_1$; $a_1 b_2, a_2 b_2, \ldots, a_n b_2$; $\ldots a_1 b_m, a_2 b_m, \ldots, a_n b_m$. (An example would be if C_A were the collective whose members are the throws of a certain die, considered in respect of turning up $1, 2, \ldots, 6$; if C_B were the collective whose members are the throws of a certain penny considered in respect of turning up heads or tails; and if R were the relation of simultaneity between a throw with the die and a throw with the coin.) The problem here is to infer the limiting frequencies of each of the $n.m$ conjunctive alternatives in the new class from the limiting frequencies of each of the n alternatives in C_A and the limiting frequencies of each of the m alternatives in C_B.

The reader may have noticed that I have spoken of forming a new *class*, and not of forming a new *collective*, by this method. The reason is that it is not necessary that a class formed in this way out of two collectives should be itself a collective. Certainly it will have one of the two defining properties of a collective, *viz*., that the frequencies with which each of the alternative possibilities is manifested by its members has a limiting value. But it need not have the other property, *viz*., randomness, *i.e.*, the indifference of these limiting frequencies to ordinal selection from the class. Now, unless the class formed by combination be itself a collective, the limiting frequencies with which the various alternatives are manifested by its members will not be 'probabilities', as defined by von Mises. He gives the following example of two collectives which are not combinable into a collective. Suppose that C_A consists of the measured values of a certain meteorological phenomenon at a certain place at 8 a.m. on successive days; suppose that C_B consists of the measured values of the same phenomenon at the same place at 8 p.m. on successive days; and suppose we make a new class each member of which is the values at 8 a.m. and 8 p.m. on the same day. It might be that at every full-moon a certain value of one causally necessitates the same value of the other. The new class would then not be a collective, and the limiting frequencies of the alternatives in it would not be probabilities as defined by von Mises.

Assuming that the two correlated collectives C_A and C_B are such that they can be combined to form a collective C_{AB}^R, there are still two different possibilities to be considered. C_A and C_B may either be or not be 'mutually

VON MISES 'WAHRSCHEINLICHKEIT, STATISTIK, UND WAHRHEIT'

independent'. Suppose that a bag is known to contain red, white, and blue counters and no others, and that two counters are drawn in immediate succession on a great many occasions and their colours noted. Let C_A be the class of 'first drawings' and let C_B be the class of 'second drawings' from this bag. If the rule of the game is that the first counter is to be replaced on each occasion before the second is drawn, then C_A and C_B are independent. If, on the other hand, the rule is that the first counter drawn is to be kept out on each occasion until after the second has been drawn and that the two are then to be replaced, then C_A and C_B are not independent. Von Mises gives a rather complicated definition of 'independence' on p. 62. It amounts to the following: Let C_A and C_B be two collectives whose terms can be correlated one-to-one. We say that C_B is 'independent of' C_A if, and only if, the following condition is fulfilled. Select in any way that you like an infinite class from C_A. Consider the terms of C_B which are correlated with the terms of this selected sub-class. Select from them, in any way that you like, an infinite sub-class. Then the limiting frequency of each alternative within this latter sub-class must be the same as the limiting frequency of the same alternative within the whole class C_B.

According to von Mises the only way in which you can tell whether two collectives are independent or not is by experiment. If they are independent, the limiting frequency for any alternative ab in the combined collective C_{AB}^R will be equal to the product of the limiting frequency for a in C_A and the limiting frequency for b in C_B. Otherwise there will not be this equality. The only way in which such a question can possible be decided is by carrying out a long enough series of observations.

Von Mises ends the lecture by working out in elaborate detail, in terms of his four processes of derivation, the simple problem in dice-throwing which the Chevalier de Méré set to Fermat, whose solution of it was the beginning of the calculus of probability.

The third Lecture is entitled 'Critique of the Foundations'. Von Mises first criticises certain alternative theories, and then considers criticisms on his own theory and proposed modifications of it. The most important points are the following.

(i) The classical definition of 'the probability of an event' originated with Laplace and has been handed down in successive mathematical textbooks ever since. He defined the 'probability of an event' as the ratio of the number of cases favourable to it to the total number of cases both favourable and unfavourable to it, all these being assumed to be '*equally possible*'. Von Mises fastens on the last proviso. He has little difficulty in showing that 'equally possible' can mean only 'equally likely to happen'. So Laplace's statement is undoubtedly circular, if taken as a *definition* of the 'probability of an event'. The only way to avoid this charge of circularity is to say that Laplace takes the notion of '*equally* probable' as indefinable, and then proceeds to define the statement that the probability of an event is so-and-so in terms of this notion.

Now anyone who takes this view will be in difficulties whenever he has to deal with a case, such as a loaded die, where the probabilities of the various alternatives are not equal. He will have to try to split up the unequally probable alternatives into disjunctions of different numbers of more fundamental alternatives all of which are equally probable; or else to admit that the theorems of the calculus of probabilities cannot be applied. Now von Mises makes the following criticisms at this point. (a) This kind of analysis, even if it can be performed, is extremely artificial in the case of loaded dice, insurance problems, etc. (b) Yet no one hesitates to apply the theorems of probability to the limiting frequencies which are found by observation in these cases. And (c) in point of fact the equal probabilities, in the case of a die which is fair, have to be established in precisely the same empirical way as the unequal probabilities in the case of a die which is loaded. In each they are simply the limiting frequencies with which the various alternatives present themselves in a collectivity of throws. If the limiting frequencies for the various alternatives $1, 2, ..., 6$ are all $1/6$ in the case of die A and are, *e.g.*, $1/21, 2/21, ..., 6/21$, respectively, in the case of die B, there is no rational ground for regarding the latter set of unequal probabilities as any less fundamental than the former set of equal probabilities.

Von Mises suggests that people have thought that equi-probability is fundamental, because they have thought that there are cases in which they could tell *a priori* that the alternatives are *equally* probable, whilst no-one imagines that he can tell *a priori* what are the probabilities of the various alternatives when they cannot be seen to be equi-probable. He is referring, of course, to the so-called 'Principle of Indifference'. He argues, quite successfully in my opinion, that in any actual case the evidence for equi-probability is always empirical, though it does not always take the form of carrying out a series of trials with the particular object under consideration. In dealing with any particular die or penny, we know that dice are generally deliberately made as 'fair' as possible, that pennies are generally made with a head and a tail and not with two heads or two tails, and so on. Again, suppose we did know *a priori* that an *accurately* cubical object, made of *perfectly* homogeneous material, would be equally likely to fall with any of its faces uppermost if *fairly* thrown. How could we possibly apply this *a priori* knowledge in any particular case? As a matter of fact we know quite well that a die is *not* a perfectly symmetrical object, since it has different numbers of spots on different faces. How can we tell, except empirically, that this difference is irrelevant to the frequency with which these variously spotted faces will fall uppermost?

(ii) The Laplaceans profess to find in Bernoulli's and Bayes's theorems, *i.e.*, in the laws of Great Numbers, a 'bridge' by which they can pass safely to and fro between their definition of 'probability' and the frequency-theory. Von Mises holds that this view is fallacious; but the point must be deferred until we consider his account of these laws.

(iii) Von Mises uses the well-known paradoxes and contradictions, which arise when the Principle of Indifference is employed to determine the probabilities of a *continuous* set of alternatives, in order to reinforce his contention that the Principle is worthless and that probabilities are always limiting frequencies based either on direct observation or postulated hypothetically and tested by observable consequences. In this connexion he criticises von Kries's '*Spielraum*'-theory of probability.

It seems to me that von Mises' criticisms on alternative theories are highly damaging; it remains to be seen what he has to say in answer to attacks on his own theory.

(i) We may defer his answer to the contention that there is a contradiction between the frequency-definition of 'probability' and the result of Bernoulli's theorem.

(ii) It may be objected that, according to von Mises, probabilities are defined as the *limits* to which observed frequencies within a class approach indefinitely near as the number of members of the class is indefinitely increased, and that nevertheless he describes them as physical properties discoverable by observation. To this the only answer that I can find is the retort that mechanics makes use of the notions of points, material particles, etc.; and that the notions of density, velocity, etc., in physics all involve proceeding to a limit and yet are determined by experiment and observation.

(iii) Two objections may be made in respect of the 'randomness' which is an essential part of von Mises' definition of a 'collective'. (a) It might be contended that the phrase 'infinite class for which there is no intrinsic rule of construction', which is what von Mises' definition of a 'collective' seems to involve, is simply meaningless. To this von Mises' answer is that the Formalist school of mathematicians need not object, provided that phrase is not self-contradictory, and that the Intuitionist school need not object, provided that a series answering to this description could be constructed by a procedure which they admit in other cases to be valid. Now the phrase has not been shown to *be* self-contradictory, though it has also not been shown *not* to be so. And Intuitionists do admit series for which the only rule of construction is to throw a die continually and note what turns up on each occasion.

(b) It might be objected that, even if the two factors in von Mises' definition of a 'collective' are severally intelligible, yet they are mutually inconsistent. It might be said that, unless there is a law connecting position of a term in a series with the alternative which it manifests, it is meaningless to talk of the frequency with which that alternative is manifested within the series as having a limiting value. Yet the condition of 'randomness' just is the condition that there is no such law. To this von Mises makes the following answer. (α) There are plenty of series which *are* given by an intrinsic rule, where, nevertheless, we *cannot* say whether the frequency of a certain alternative has a limiting value or not. (An example is the following. Suppose you take the series of digits in the endless decimal which expresses the square-root of π, and substitute a 0 for each even digit and a 1 for each odd digit. The series is constructed according to a rule; but there is no answer to the question whether the frequency with which 1's occur in it has a limiting value.) This appears to me to be interesting, but quite irrelevant to the objection under discussion. (β) He says that, unless there is something in the description of a series which positively *excludes* the possibility of the frequency of an alternative in it having a limit, you are at liberty to suppose that there is such a limit and to work out the consequences. (γ) If it is objected that this reduces the whole calculus to a game, he points to the practical applications of the theory in physics and social statistics.

VON MISES 'WAHRSCHEINLICHKEIT, STATISTIK, UND WAHRHEIT'

I think it must be admitted that the objections which we have been considering are highly plausible, and that von Mises' answers to them are not very convincing. But I think that we can go further. These objections may be called 'logical', in the sense that they raise doubts as to whether any clear meaning can be attached to the statement that there are 'collectives' and 'probabilities' as defined by von Mises. But, even if these logical difficulties could be removed, a serious epistemological question would remain. How are we justified in passing from the empirical premise that the frequency with which a certain die has fallen with 6 uppermost in the N times which, so far as we know, it has been thrown is so-and-so, to the conclusion that, if it were thrown infinitely many times, the frequency would approach indefinitely near to the limiting value so-and-so? Again, how can we establish empirically the very sweeping universal negative proposition that there is *no* way of selecting an infinite sub-class from the original class of throws which would have a different limiting frequency for the same alternative? If we have any rational ground for believing such conclusions on such evidence, must it not involve principles of 'probability' in some important sense of 'probability' not contemplated by von Mises? This would not necessarily be any objection to von Mises' definition; for he is admittedly confining his attention to 'probability' in the sense in which it can be measured and made the subject of a calculus. But it would show that we should have no reason to believe any propositions about probability, in his sense, unless there are logical principles of probability, in another sense.

The rest of Lecture III is devoted to writers who agree in the main with von Mises but propose a less rigid condition in defining collectives than that of complete randomness. The least rigid of these suggested conditions is that the series must be 'Bernoullian'. Suppose that p is the probability of a certain alternative being manifested by a term in the series, and suppose that we take as the terms of a new series the first n, the second n,... and so on, terms of the old series. Then the Bernoullian condition is that the probability of any term in the new series being any particular ordered sequence of r occurrences and $n-r$ non-occurrences of the given alternative must be $p^r(1-p)^{n-r}$ for all values of n and r. Other writers, such as Popper and Reichenbach, have proposed a more rigid condition, which includes the Bernoullian condition and another besides. Von Mises claims to show that series can be constructed which answer to these con-

195

ditions and yet have limiting frequencies for the occurrence of certain alternatives which no-one in his senses would admit to be the *probabilities* of those alternatives. Hence a more rigid condition is needed in order to demarcate collectives whose limiting frequencies shall be what are commonly taken as the probabilities of such and such alternatives. He mentions the American mathematician Copeland as one who has come nearest to defining conditions which are sufficient and yet are less sweeping than his own condition of complete randomness.

Finally, on pp. 120 to 122, von Mises gives a sketch of the work of the mathematician Dörge, who has tried to construct an axiomatic theory on von Mises' lines which shall avoid the criticisms brought against the theory in its original form. This looks very interesting, but it is too technical to be summarised here.

We can now pass to the fourth Lecture, which deals with the Laws of Great Numbers, *i.e.*, with Bernoulli's, Poisson's, and Bayes's theorems, and with later extensions and polishings of these. Von Mises' discussion of these questions seems to me to be extremely valuable and illuminating.

Let us take Bernoulli's theorem and Bayes's theorem as typical, since the former is simple to state and the latter is, in a certain sense, the 'inverse' of it. We will begin with Bernoulli's theorem. I think that the essential points in von Mises' discussion of it may be stated as follows.

(i) Whatever meaning we may attach to the word 'probability', both the premises and the conclusion of Bernoulli's theorem are in terms of probability.

(ii) The correct statement of the theorem is as follows. Suppose that the probability of a certain alternative being realised on any one occasion of a certain kind is p. (Take, *e.g.*, the probability of throwing a 6 in any one throw with a certain die.) Consider a set of n such occasions; *e.g.*, n successive throws with this die. Let ε be any fraction, *e.g.*, one-millionth. Let $\pi_{n,\varepsilon}$ be the probability that this alternative will be manifested not less than $pn-n\varepsilon$ times and not more than $pn+n\varepsilon$ times in such a set of n occasions. Then, no matter how small ε may be, the probability $\pi_{n,\varepsilon}$ will approach indefinitely near to 1 as n is indefinitely increased.

(iii) We must now interpret this proposition when 'probability' is defined in terms of limiting frequency. I shall state it in my own way, but I shall be giving what is in fact von Mises' interpretation of it. Consider, *e.g.*, a series each member of which is *a single throw* with a certain die.

Let N be the total number of times it has been thrown, and let N(6) be the total number of these which have turned up 6. It is assumed that the ratio N(6)/N approaches indefinitely near to a certain limit p as N is indefinitely increased. And it is assumed that this series is 'random'. Now consider a new series each term of which is a *set of n throws* with the same die. Let ε be any fraction, *e.g.*, one-millionth. Let N' be the total number of such *sets* that have occurred, and let $N'(pn \pm n\varepsilon)$ be the number of such sets which contain not less than $pn - n\varepsilon$ and not more than $pn + n\varepsilon$ 6's in each. Then (a) the new series is 'random'. (b) The ratio $N'(pn \pm n\varepsilon)/N'$ approaches indefinitely near to a certain limiting value $\pi_{n,\varepsilon}$ as N' is indefinitely increased. And (c) no matter how small ε may be, this limiting ratio $\pi_{n,\varepsilon}$ will approach indefinitely near to 1 as n, the number of terms in each set, is indefinitely increased. This conclusion may be summed up more colloquially as follows. However small ε may be, if you increase the number of *terms in each set* and the *number of sets* sufficiently, an overwhelming majority of the sets will contain a proportion of 6's which differs from p by less than ε.

(iv) It is sometimes objected that, if the frequency-theory of probability were true, Bernoulli's theorem would consist in laboriously proving what is already asserted in the premise that the probability of a certain alternative being realised on any one occasion is p. It is quite evident from the interpretation of the theorem given above that this objection is mistaken.

(v) On the other hand, it is sometimes objected that the frequency-theory assumes something to be *certain* which the Bernoulli theorem proves to be only *very probable*. In the case of a die, *e.g.*, the frequency-theory assumes that the ratio N(6)/N has a certain exact limiting value p when N is indefinitely increased. But the Bernoulli theorem, it is alleged, shows that we have no right to assert more than that N(6)/N is very unlikely to differ by more than a certain pre-assigned small amount from p if N be made large enough. A glance at the accurate statement of the theorem above will show that this objection is invalid. The conclusion of the theorem, in our notation, is not about the limiting value of N(6)/N in the original series of *single throws* as N is indefinitely increased. It is about the limiting value of $N'(pn \pm n\varepsilon)/N'$ in the series of *sets of n throws* when both n and N' are indefinitely increased.

(vi) The notion that Bernoulli's theorem could act as a 'bridge' between 'probability' in the Laplacean sense and 'probability' in the frequency sense is a complete delusion. In whatever sense 'probability' is used in the premises it must be used in that sense in the conclusion. Let us take a concrete example to illustrate this. Bernoulli's theorem shows that, if the probability in the Laplacean sense of throwing a head with a certain coin is $\frac{1}{2}$, then the probability in the Laplacean sense of getting between 49 per cent and 51 per cent of heads in a set of 10,000 throws with this coin is approximately .95. It also shows that the probability in the Laplacean sense of getting between 49 per cent and 51 per cent of heads in a set of 100 throws with this coin is approximately .16. Now in cases like the first, where the Laplacean probability is nearly 1, there is a strong tendency to pass surreptitiously from the Laplacean probability to assertions about limiting frequency. There is a strong tendency to state the conclusion in the form that in almost all sets of 10,000 throws the percentage of heads will fall between 49 and 51. But would a Laplacean be prepared to make a similar identification of Laplacean probability with limiting frequency in the second case, and to say that in 16 per cent of sets of 100 throws the percentage of heads will fall between 49 and 51? If it is justifiable to identify *high* Laplacean probabilities with limiting frequencies of nearly 100 per cent, surely it must be equally justifiable to identify any lower Laplacean probability with a correspondingly lower limiting frequency. The plain fact is this. You cannot legitimately draw any conclusion about the limiting frequency with which a *certain proportion of heads* will occur in a series of *sets of n throws* unless you start with a premise about the limiting frequency with which *a head* will occur in a series of *single throws*. And, beside this premise, you will need the further premise that the occurrence of heads is 'randomly distributed' in the original series of single throws, in the sense explained above.

Having, as I hope, made von Mises' position about Bernoulli's theorem and its relation to the frequency theory quite clear, I can deal much more briefly with Bayes's theorem. I shall again state von Mises' view in my own way. In order to be as concrete as possible I will again talk in terms of dice.

Suppose you have a set of N dice, each of which has been thrown n times and has given the *same* number $n(6)$ of sixes. Let $N(p)$ be the number of these dice which, if thrown an indefinitely large number of times, would turn up 6 with the limiting frequency p. (Of course p is a proper fraction capable of having any value from 0 to 1 inclusive.) Then (a) for every possible value of p the corresponding ratio $N(p)/N$ has a characteristic limiting value as N is indefinitely increased. (b) Let ε be any fraction, e.g., one-millionth, and let $N[n(6)/n \pm \varepsilon]$ be the number of these dice which, if thrown for an indefinitely large number of times, would turn up 6 with a limiting frequency not less than $n(6)/n - \varepsilon$ and not greater than $n(6)/n + \varepsilon$. Then the ratio of $N[n(6)/n \pm \varepsilon]$ to N approaches indefinitely near to 1 as limit when both n and N are indefinitely increased, no matter how small ε may be. The conclusion may be summed up more colloquially as follows. Suppose you have a very large number of dice, each of which has been thrown a great many times and has given the same proportion of 6's. Let ε be any fraction. Then, if only the dice be numerous enough and you throw each of them long enough, an overwhelming majority of them will give 6's in a proportion which differs by less than ε from the observed proportion, no matter how small ε may be.

It is obvious that this theorem is of the utmost importance for the practical application of the frequency theory of probability. For the essential point of it is the following. It enables you to start with the *observed* frequencies in a number of similar series, and to conclude that the *limiting* frequencies in the great majority of these series differ very little from the observed frequencies.

Lecture IV concludes with a fascinating account of the extensions of Bernoulli's and Bayes's theorems which have been made in recent years by Polya and Cantelli, and with an introduction to the notion of Statistical Functions.

I shall touch very lightly on the two remaining lectures, although they are of extreme interest. In Lecture V von Mises explains and deals with Marbe's problem of the expectant father who hopes that his child will be a boy and studies the recent birth-statistics; with Polya's treatment of the statistics of epidemics: and with Lexis's notion of normal, sub-normal, and super-normal dispersion. The fundamental problem of statistics, according to von Mises, is to discover whether a given set of observations can be regarded either (a) as a finite part of a certain collective, or (b) if not, can be regarded as following by certain assignable processes from certain collectives. He compares the whole procedure to Kepler's observations leading first to Newton's laws of planetary motion and these leading in turn to the calculation of the actual complex and not truly elliptical paths of the planets. The lecture ends with a discussion of the theory of errors of observation, illustrated by the device known as Galton's Board.

The sixth and last lecture deals with the applications of probability in physics. It treats of the classical kinetic theory of gases; the theory of Brownian movement; the theory of radio-active discharge; the more recent developments of gas-theory by Einstein, Bose, and Fermi; and the Uncertainty Principle in quantum mechanics.

It would be difficult to recommend this book too highly. It is written with admirable clearness; it presupposes no advanced mathematical knowledge; it is full of the most interesting examples; and it provides at intervals admirable summaries of the argument and the conclusions. It is very much to be hoped that it will be translated into English.[1]

NOTE

[1] *Editor's note:* This has in fact happened after Professor Broad's critical notice was published. A translation of the second German edition (the one reviewed by Broad) was published by William Hodge & Co., London, in 1939, under the title *Probability, Statistics and Truth*. A considerably modified second edition of this translation, mainly following the third German edition (1951), was published by George Allen and Unwin Ltd., London, in 1957.

Some terms used in Professor Broad's review have been changed so as to conform with the terminology adopted in these translations.

CRITICAL NOTICE:

Probability and Induction. By WM. KNEALE. Oxford: Clarendon Press. Pp. viii+264.

This very able and interesting book is based on the lectures given in Oxford by Mr. Kneale up to the outbreak of the second world war, and has been prepared by him for the press in the scanty leisure which he has enjoyed since that war changed from 'hot' to 'cold'. It forms an admirable general introduction to the philosophy of the two inter-connected subjects named in its title, but what makes it particularly exciting is certain special doctrines on fundamental questions which Mr. Kneale asserts and defends. These are in conflict with certain philosophical principles or prejudices which are at the moment fashionable and almost orthodox among Mr. Kneale's contemporaries and juniors in this country and the United States. These parts of the book are likely to lead to much valuable discussion. It is a very happy circumstance that doctrines which are at the moment unfashionable should be put forward by a man like Mr. Kneale, who is fully aware of the strength and the weaknesses of the current orthodoxy, and whom no-one in his senses can afford to dismiss as a negligible 'back-number'.

The doctrines to which I refer are the following. Mr. Kneale distinguishes between matters of fact and what the calls 'Principles of Modality'. He rejects the view that all statements which ostensibly record principles of modality are really statements about language couched in a misleading form. He holds that, if there are laws of nature, they are all principles of necessity, although none of them can be known *a priori*. Lastly, he holds that what he calls 'Probability Rules', *i.e.*, propositions of the form 'The probability of an instance of α being an instance of β is p', are also principles of modality, which cannot be known *a priori* but can be reasonably conjectured inductively on the basis of statistics. According to

Mr. Kneale, the laws of logic, of phenomenology, and of nature (which are all fundamentally of the same kind), leave open a certain range of possibility for anything which is an instance of α, and they leave open a certain narrower range of possibility for anything which is an instance of $\alpha\beta$. What a probability rule asserts is that the latter bears a certain proportion to the former.

I shall begin by giving a rough general sketch of the contents of the book as a whole, and shall then try to expound in greater detail (so far as I understand them) these characteristic doctrines of Mr. Kneale's and his reasons for them and against alternatives to them.

I. GENERAL OUTLINE OF THE CONTENTS

The book is divided into four Parts. The first, which is introductory, treats of *Knowledge and Belief*. The second, entitled *The Traditional Problem of Induction*, is concerned with all the main philosophical problems of induction in so far as that process is used to establish *laws*, as distinct from probability-rules. The third, entitled *The Theory of Chances*, discusses the fundamental notions and theorems of the calculus of probability, and considers whether these are relevant to the logic of induction. The answer to the latter question is negative; and the fourth Part, entitled *The Probability of Inductive Science*, deals with the question *whether*, and, if so, *in what sense*, recognized inductive procedures give more or less 'probability' to statements of law and to probability-rules.

1. *Knowledge and belief*

Mr. Kneale's conclusions may be summarized as follows. He starts with the antithesis between 'knowing p' and 'believing p'. He holds, in opposition to some distinguished epistemologists, that 'knowing' is used in an occurrent sense, and not *only* in a dispositional sense. (*Cf., e.g.*, "When it began to pour with rain while I was out walking this afternoon I *knew* that I should get wet through".) He has not met with any satisfactory analysis of 'knowing p', in the occurrent sense, and so he takes it provisionally as unanalyzable. It is equivalent to 'noticing that p' or 'realizing that p'.

The phrase 'believing p' covers two different cases, which may be described as 'taking p for granted' and 'opining p'. The former consists in acting as if one knew *p* when one does not know it. To say that *A* opines *p* with a certain degree of rational confidence means that (i) *A knows* certain other propositions *q*, which in fact probabilify *p* to the degree in question, and (ii) *A knows* that *q* probabilifies *p* to that degree. Opining may be irrational. This covers two cases. *A* may take for granted (instead of knowing) some or all of the evidence for *p*; or he may take for granted (instead of knowing) that the evidence probabilifies *p* to the degree in question. Neither failure in rationality necessarily leads to false belief.

Probabilification of one proposition to a certain degree by certain other propositions is a purely objective fact. It has certain analogies to the necessitation of one proposition by another, and certain unlikenesses to it.

We talk of the probability of throwing a six with a fair die, and we also say that induction establishes certain laws and certain probability-rules with high probability. Mr. Kneale holds that the word probability is used in different senses in these two applications. But it is not just a single word with several totally disconnected meanings, like the word 'plot', *e.g.* There are real and important analogies between its various applications. A most important common feature is that it is reasonable to take as a basis for action any proposition which is highly probable, in the appropriate sense, on the evidence available to one. Any satisfactory analysis of 'probability' must enable us to see why this is so.

2. The traditional problem of induction

Taking induction for the present as a process by which universal propositions are established, Mr. Kneale points out that the word has been used to cover four different processes, each of which leads to a different kind of universal proposition. These processes may be described as *Summary*, *Intuitive*, *Mathematical*, and *Ampliative* induction.

Summary Induction establishes propositions of the form All S is P, where the description 'S' is such that, from the nature of the case, it can apply only to a finite number of instances, and where it is in principle possible to know that one has exhausted the whole set. An example would be: All the chairs in this room on Christmas Day 1946 had red seats. Mr. Kneale points out that such a statement is equivalent to: No part of this room during the period in question was occupied by a chair with a seat which was not red. This is different in kind from such a proposition as: All men are mortal. Summary Induction is a deductive argument, though it cannot be reduced to a syllogism in the technical sense. One premiss has to be what might be called an 'exhaustive' proposition; *e.g.*, This, that, and the other sub-region together make up the space in this room.

The result of Intuitive Induction is knowledge of what Mr. Kneale calls 'Principles of Modality', *i.e.*, of compatibility or incompatibility between characteristics. These are essentially universal and necessary. An example would be: No surface could be red and green all over at the same time, but a surface could be at once red and hot all over. It is a characteristic doctrine of Mr. Kneale's that such propositions are *not* merely linguistic. His arguments on this point will be considered later. We may sum up Mr. Kneale's account of intuitive induction by saying that he considers it to be a valid intellectual process, but not a form of reasoning. What experience does here is to provide *instances*, not premisses.

Mathematical Induction, or argument by recurrence, establishes propositions of the form: All numbers have the property p. Such propositions are necessary, but they differ in kind from principles of modality which are established by intuitive induction. After considering various alternative views as to the nature of propositions about all numbers, Mr. Kneale puts forward the following account of them. To say, *e.g.*, that 2 is a number, is to say that '2' is a *recurrence symbol*, *i.e.*, that it signifies, not an individual nor a character of an individual or a group, but a certain feature, *viz.*, a recurrence in the structure of such facts as are expressed by sentences like 'There are 2 tables in this room'. To say that all numbers have the property p is equivalent to saying: '1 has the property p, and, if c has it, then $c+1$ has it'. Thus, such propositions depend on the fact that the whole nature of numbers is to form a sequence generated by arithmetical addition.

Mr. Kneale argues that all proofs of universal propositions about numbers involve mathematical induction. For propositions about other kinds of number are reducible to complicated statements about integers, and all proofs of universal propositions about integers depend on mathematical induction. Proofs which seem *prima facie* to be independent of this process involve the principles of algebra, *e.g.*, the associative law, and these can be proved only by recurrence.

Ampliative Induction is concerned to establish natural laws and probability-rules. For the present we will confine our attention to the former. A law of nature is a proposition of the form: All S is P, where the description 'S' applies to a potentially unlimited class of individual existents.

Mr. Kneale distinguishes the following four types of law. (i) *Uniform associations of attributes.* There are the laws which are involved in the existence of those groups of associated properties which mark out natural kinds. (ii) *Uniformities of development in natural processes.* Examples are found in the course of development of an embryo, of a chemical reaction, and so on. The Second Law of Thermodynamics is an advanced instance. (iii) *Laws of functional relationship.* An example would be the gas-law $PV = RT$. Such laws require that there shall be a uniform relationship between values of the several variables, and that this shall be expressible in a formula of pure mathematics. (iv) *Numerical natural constants.* An example would be the law that gold melts at such and such a temperature. (It will be noted that each of the last three kinds of law involves a reference to natural kinds, *e.g.*, embryos of *mammals*, instances of *gases*, bits of *gold*.)

Mr. Kneale gives an interesting historical account of the senses in which the word 'cause' was used by Aristotle, by Bacon, and by Hume and his continuator Mill. In this connexion he gives a critical account of the eliminative methods proposed by Bacon and by Mill for discovering 'the cause' or 'the effect' of a given phenomenon. His general conclusion is that philosophers have tended to exaggerate the importance of the notion of cause in science. It is a vague notion, useful enough in some departments of practical life, but incapable of being made unambiguous and precise. When one tries, as Hume and Mill did, to tie it down to the notion of 'antecedent cause', it develops ambiguities and difficulties; and to describe science as a search for causes and causal laws, in this sense, is to give an inadequate and misleading account of the procedure of the more advanced sciences.

The most important section of this Part is concerned with the logical or ontological nature of laws. I shall expound Mr. Kneale's views and his reasons more fully later. For the present it will suffice to say that he considers and rejects the following views about natural laws, *viz.*, (i) that they are analogous to the restricted universals established by summary induction, (ii) that they are *facts* (as opposed to *principles*) of unrestricted generality, *i.e.*, the *de facto* regularity analysis, and (iii) that they are merely

205

regulative prescriptions. Every alternative has its difficulties, but in the end Mr. Kneale accepts and defends the view that laws are *principles of modality*, i.e., are of the same nature as the propositions which are established by intuitive induction, although for reasons which he gives, no law can be established in that way. This alternative, he says, is at any rate 'not entirely hopeless', whilst all the others are so. It is 'the only account of laws which makes sense'.

There are hosts of alleged laws for which there is good inductive evidence, and serious science begins when we try to correlate and explain them. Such explanation may take two forms. (i) We may try to show that a large number of these laws follow logically from one or more others which have themselves been established by direct induction. (ii) We may try to show that they are entailed by one or more propositions which have not been, and from the nature of the case *could never be*, established by direct induction. Mr. Kneale calls the latter 'explanation by means of *Transcendent Hypothesis*'. An example of a transcendent hypothesis is the atomic theory or the wave-theory of light.

The peculiarity of a transcendent hypothesis is that the things and processes in terms of which it is formulated *could not* conceivably be perceived by the senses, and therefore, strictly speaking, could not be imagined either. It is plain that such hypotheses raise certain philosophical questions. Mr. Kneale's main answers are as follows:

(i) Although we cannot imagine a transcendent entity, we conceive it as having a certain definite logical or mathematical *structure* embodied in a content which we cannot even conjecture. (ii) Any statement about a perceptual object, *e.g.*, a table, can be translated into statements about transcendent objects, *e.g.*, electrons and protons; but there are many significant statements about transcendent objects, *e.g.*, about what happens inside an atom, which cannot be translated into statements about perceptual objects. (iii) Some of these non-translatable statements about transcendent objects are essential if the hypothesis is to explain known laws about perceptual objects and to suggest others which may be tested experimentally. (iv) For the above reasons Mr. Kneale rejects the view that statements about transcendent objects are merely a new and mathematically more handy terminology for talking about perceptual objects and their laws. He thinks that it would be unintelligible, on that view, that a transcendent hypothesis should enable one to infer laws about perceptual objects which had not as yet been established by direct induction. I do not find this argument very convincing. I suppose, *e.g.*, that the difference between the heliocentric and the geocentric descriptions of the planetary motions is merely a difference in the frame of reference adopted. Yet it is almost inconceivable that Kepler's laws of planetary motion would have been discovered unless the heliocentric description had been substituted for the geocentric; and, unless they had been, it is almost inconceivable that the law of gravitation would have been discovered.

Mr. Kneale uses the term 'secondary induction' for the kind of reasoning by which a transcendent hypothesis, as distinct from an ordinary law about perceptual objects, is experimentally verified or refuted.

Suppose that a hypothesis H entails a number of laws, $L_1, L_2, ...$, for each of which there is direct inductive evidence, $e_1, e_2, ...$, respectively. Then each of these laws is supported indirectly by the direct evidence for all the others. For e_r, in supporting L_r, supports the hypothesis H, which entails L_r. And, in supporting H, it indirectly supports any other law, L_s, which is entailed by H. This is called by Mr. Kneale 'consilience of primary inductions'. It plays an important part in every advanced science.

The last topic dealt with in this Part is the relative importance of confirmation and elimination in induction. Mr. Kneale points out that elimination can lead to no positive conclusion unless it can be combined with *some* affirmative universal premiss. Now, even if some general principle of determinism could be formulated and were found to be self-evident, it would be far too abstract to serve as a useful premiss in an eliminative argument. In fact when scientists use such arguments they employ fairly concrete positive premisses, such as, *e.g.*, the proposition that all samples of a pure chemical substance have the same melting-point. Now these have to be established in the end by positive confirmatory inductive argument. So the fundamental problem of induction is confirmation by positive instances, and not elimination by negative instances.

3. *The theory of chances*

Mr. Kneale defines a 'probability-rule' as a statement of the form: The probability of an instance of α being β is so-and-so. He symbolizes such a rule by the formula $P(\alpha, \beta) = p$. The calculus of chances is described as the procedure for deriving new probabilities from others which are given.

Mr. Kneale states the axioms and develops the theorems. All this is well done, but it raises no points of special interest. As might be expected, Mr. Kneale is under no illusions about the nature of Bernoulli's limit theorem, which he proves without using the differential calculus. He points out that it, like all theorems in the calculus of probability, merely derives one probability from another. On the other hand, it is a *necessary* proposition, and it is absurd to treat it as a law of nature which might be supported or refuted by experiments with coins or dice. It will be worth while, in this context, just to mention the notation which Mr. Kneale introduces for stating and proving theorems about the probability of a set of individuals having a certain composition. He uses the symbol $P(^{\alpha}\sigma_n, {}^{\beta}\kappa_m)$ to denote the probability that a set of n instances of α contains exactly m instances of β. He uses a similar symbol, with ${}^{\beta}\rho_p$ substituted for ${}^{\beta}\kappa_m$, to denote the probability that such a set contains a proportion p of instances of β.

I think that these symbols betray an inadequacy which was already latent in the notation $P(\alpha, \beta) = p$. What Mr. Kneale in fact wants to symbolize is the probability that a set of n instances of α will contain exactly m instances of β, *given that* it is selected under certain conditions which might be called 'Bernoullian' and *given that* the probability of an instance of α, so selected, being β is p. He has to state all this separately in words, and is unable to embody these conditions in his symbols. Yet, in the absence of some explicit reference to the first of these conditions, the symbol $P(^{\alpha}\sigma_n, {}^{\beta}\kappa_n)$ has no definite *meaning*; and, in the absence of some explicit reference to the second of them, it has no definite *algebraical form*, such as, e.g., ${}^{n}C_m p^m (1-p)^{n-m}$.

Before leaving this part of my exposition I will mention that Mr. Kneale states and proves two interesting theorems of Poincaré's, one about the results of spinning a roulette-wheel, and the other about those of repeatedly shuffling a pair of cards. These he calls 'equalization theorems'.

The philosophically interesting contents of this Part begin in §32, where Mr. Kneale starts to investigate the Frequency Theory of the meaning of probability rules. He takes von Mises' form of this theory as the best available for discussion. This defines $P(\alpha, \beta)$ as the limit which the proportion of instances of β in a succession of instances of α approaches as the number of terms increases indefinitely, provided

A CRITICAL NOTICE OF 'PROBABILITY AND INDUCTION'

that the succession is of the kind which von Mises calls a 'collective'. This condition is that, if any endless sub-class be selected from the original succession, in accordance with any rule of place-selection, no matter how fantastic, the limiting proportion of β's in it will be the same as that in the original succession.

There are several well-known and obvious *prima facie* objections to this definition, and von Mises or his followers have attempted to answer them. Mr. Kneale gives a clear and fair account of these objections and the attempted answers. We may pass over this and consider what he has to say on his own account.

(i) The frequentists have often defended their notion of limiting frequencies by alleging that they are analogous to certain limiting notions which are constantly used in science, and to which no-one objects. Mr. Kneale complains that it is not clear what precisely they have in mind here. Is it the ideal figures of pure geometry in contrast with the imperfect straight lines, circles, etc., which occur naturally or can be constructed artificially? Or is it such notions as frictionless fluids, perfect gases, and so on? I should have suspected that it was neither of these, but rather the notions which are expressed by such phrases as 'density-at-a-point', 'velocity-at-an-instant', and so on. However this may be, Mr. Kneale objects that pure geometry is not a natural science and is quite indifferent to whether there are perfect circles, etc., in nature. Again, physicists know very well that there are no frictionless fluids or perfect gases. But the frequentists *define* such statements as '$P(\alpha, \beta) = p$' in terms of collectives and limiting frequencies, and they believe that many probability statements apply within the actual world. Therefore they cannot afford to be indifferent to the question whether there actually are collectives with limiting frequencies, still less can they afford to admit that there are none.

(ii) The definition of a 'collective' involves the notion of *laws* in the strict sense, *i.e.*, propositions of the form: Every instance of S (where the extension of S is potentially unlimited) is P. But these laws are of a very odd kind, and it is very difficult to see why anyone should think he has good reason to accept them. For they are of the form: *Every* infinite selection made by *any* rule of place-selection from the endless succession of α's contains the same limiting proportion of β's as the original succession.

(iii) The notion of a collective of α's in which the limiting proportion of β's is 1 covers two cases which common-sense sharply distinguishes. One is that of law, *viz.*, Every instance of α is β. The other is the case where, although the limiting ratio is 1, yet there are many (it may be infinitely many) instances of α which are not β. If the frequentist thinks that he can get rid of the notion of law and reduce all instances of unitary probability to the second heading, it is plain from what has been said above that he is mistaken.

(iv) Consider, *e.g.*, the following application of Bernoulli's Theorem. If the chance of throwing a 5 with a certain die is 1/6, then there is a very high probability that the percentage of 5's in a set of 1000 throws with that die is in the near neighbourhood of 16.66 per cent. Now let us interpret this in terms of the frequency theory. It will run as follows. If in an endless succession of *single throws* with this die the limiting ratio of the number of 5's to the number of throws is 1/6, then in an endless succession of *sets of 1000 throws* with it the limiting ratio of the number of such sets with about 16.66 per cent of 5's in each of them to the number of such sets is not far short of 1. Now would a knowledge of the antecedent proposition about the properties of an endless succession of *single throws* give you any good reason to bet on a non-5 rather than a 5 at *the next throw?* And would a knowledge of the consequent proposition about the properties of an endless succession of *sets of 1000 throws* give you any good reason to bet on a percentage of 5's near to 16.66 per cent in *the next set of 1000 throws?* The answer in both cases seems plainly to be No. Yet a satisfactory analysis of probability-rules ought to account for the fact that we think it reasonable to use them as guides to action in making decisions about particular cases and particular sets of many cases.

For such reasons as these Mr. Kneale rejects the frequency theory of the meaning of the probability-rules.

Mr. Kneale approaches his own theory of the meaning of probability rules by way of a discussion on the notions of Equiprobability and Indifference. He rejects, on the usual and quite conclusive grounds, the Principle of Indifference, *i.e.*, that alternatives are equally probable relative to a person's state of information if he knows of no reason for accepting one rather than another of them. But the fact that this principle gives no satisfactory *criterion* for judging whether several alternatives are equiprobable does not show that it is a mistake to *define* the measure of a probability in terms of equiprobable alternatives.

In developing his own theory Mr. Kneale begins with the case of a characteristic which has a finite range of application, *e.g.*, the concept of undergraduate of Oxford in 1949. To say that two alternatives under such a concept are 'equipossible' is equivalent to saying that either (i) both are *ultimate*, *i.e.*, non-disjunctive, relative to that concept, or (ii) that each consists of a disjunction of the same number of ultimate alternatives under it. An example under the first heading would be the alternatives of being this or that Oxford undergraduate in 1949. An example of alternatives which are not equipossible, relative to the sizes of the two colleges, are those of being an undergraduate of Christ Church or an undergraduate of Merton. If α is a characteristic with restricted application, the measure of $P(\alpha, \beta)$ is simply the ratio of the number of ultimate possibilities under [being-an-instance-of-$\alpha\beta$] to the number of ultimate possibilities under [being-an-instance-of-α]. *E.g.*, the chance that an Oxford undergraduate in 1949 will be an undergraduate of Christ Church is simply the ratio of the number of Christ Church undergraduates to the number of Oxford undergraduates in that year.

Mr. Kneale contrasts this with the indifference theory as follows. On his theory, in order that alternatives may be equipossible they must *be* indifferent in a certain way in relation to the characteristic under which they fall, whether this fact is known or believed or not. On the indifference theory the *person* who makes the judgment of equipossibility must be indifferent in a certain way in *his attitudes* towards them.

We can pass now to Mr. Kneale's account of the much more difficult and important case where α is a characteristic which applies to a potentially unlimited class of individuals, *e.g.*, the property of being a throw with a certain die. This seems to me to be much the most difficult part of the book, and I can only state in my own way what I believe to be Mr. Kneale's doctrine.

I shall begin by introducing the term 'specialization of a characteristic'. To be red is a specialization of being coloured; it may be called a 'determinate' specialization. To be a cat is a specialization of being a mammal; it may be called 'specific' specialization. To be red and round is a specialization of being red (and equally, of course, of being round); it may be called a 'conjunctive' specialization. Any characteristic A can be conjunctively specialized by conjoining it with any other characteristic B which A neither entails nor excludes. Similarly AB can be further conjunctively specialized by conjoining it with C, provided that it neither entails nor excludes C. (We must remember, in this connexion, that there is for Mr. Kneale no difference in principle between *nomic* entailment or exclusion, e.g., water cannot flow uphill, and entailment or exclusion of the phenomenological or logical kind, e.g., a surface cannot simultaneously be red and green all over). Starting with any generic characteristic, we can think of it as first being specialized specifically till we come to the notions of the various lowest species under the genus. Then we can think of each conjunct in the notion of each lowest species being specialized by becoming perfectly determinate in every possible way. Finally, we can think of each perfectly determinate specialization of each such lowest specific specialization being conjunctively specialized by combining it conjunctively with every other characteristic which it neither entails nor excludes. In this way we conceive of a set of *ultimate* specializations of the original characteristic. This, if I am not mistaken, is what Mr. Kneale means by the *Range* of a characteristic. Any possible individual instance of a characteristic must be an instance of *one* and *only* one of the ultimate possibilities in its range; and any two individual instances of it must be instances of *different* ultimate possibilities in its range.

Now at a certain stage in the descending hierarchy of increasingly specialized alternatives under a characteristic there will be alternatives which are *completely* specialized both specifically and determinately and can therefore be further specialized *only* conjunctively.

If I understand him aright, Mr. Kneale calls any such alternative a 'Primary' alternative. Now suppose that $\alpha_1, \alpha_2, ..., \alpha_r, ...$, are a set of mutually exclusive and collectively exhaustive *primary* alternative specializations of α. Since each is primary, any further specialization of any of them, e.g., of α_r, must be of the form $\alpha_r \theta$, where θ is a characteristic which is neither entailed nor excluded by α_r, or, as we may say for shortness, α_r is 'contingent to' θ. Suppose now that it were the case that every characteristic to which *any* of the alternatives $\alpha_1, \alpha_2, ...$ is contingent is a characteristic to which *all* of them are contingent. Then it is plain that to every specialization of any of these alternatives there would correspond one and only one specialization of each of the others. For any specialization of α_r must be of the form $\alpha_r \theta$ (since α is primary), where α_r is contingent to θ. But if α_r is contingent to θ, then any other alternative in the set, e.g., α_s, will also be contingent to θ, by hypothesis. Therefore there would be a specialization $\alpha_s\theta$ of α_s, corresponding to the specialization $\alpha_r\theta$ of α_r. Plainly there could not be more than one. And, since all the alternatives in the set are primary, none of them can have any specializations which are not of this conjunctive form. It follows that any set of alternatives under α, answering to the above conditions, would subdivide the range of α into sub-ranges, each of which covers exactly the same number of ultimate specializations of α. Accordingly, Mr. Kneale gives the name '*Primary* set of equipossible alternatives under α' to any set of mutually exclusive and collectively exhaustive primary alternatives under α, which are such that *all* are contingent to any characteristic to which *any* is contingent. Given a set of *primary* equipossible alternatives, it is of course easy to form sets of equipossible alternatives which are not primary, *viz.*, by taking as the new alternatives disjunctions of equal numbers of the old ones without overlapping, *e.g.*, $\alpha_1 \lor \alpha_2, \alpha_3 \lor \alpha_4, \ldots$.

So far we have confined our attention to the range of a single characteristic α. But, if we wish to define $P(\alpha, \beta)$, we must now introduce a reference to β. The next stage is this. Suppose there is a primary set of equipossible alternative specializations of α, such that each of them either entails or excludes β. (In general some would entail it, and the rest would exclude it.) Now, if α_r entails β or if it excludes β, it is plain that the conjunction of α_r with any other characteristic θ will also entail or exclude β, as the case may be. Thus we might say that θ in the alternative $\alpha_r\theta$ is 'superfluous' in respect of its entailing or excluding β. If there is a set of equipossible alternative primary specializations of α, each of which either entails or excludes β, it is plain that there must be such a set composed of alternatives which are *minimal* in this respect, *i.e.*, which contain nothing superfluous to entailing or to excluding β, as the case may be. If I understand Mr. Kneale aright, he gives the name '*Principal* set of alternatives under α with respect to β' to a primary set of equipossible alternatives under α, each of which either entails or excludes β, and each of which is *minimal* in that respect.

At length we come to Mr. Kneale's account of the *meaning* of the statement '$P(\alpha, \beta) = p$'. If I am not mistaken, it is as follows. The *meaning* is the same in all applications, *viz.*, the ratio of the *measure* of the range of $\alpha\beta$ to the *measure* of the range of α. But in different types of application the ranges have to be measured in characteristically different ways. (i) If α determines a *closed* class, *e.g.*, contemporary Oxford undergraduates, then the measure of the range is simply the number of individuals in the class. (ii) If α determines an open *class*, *e.g.*, possible throws with a certain die, the first move is to introduce the notion of a primary set of equipossible alternative specializations of α, each of which either entails or excludes β. Two possibilities then arise. (a) Although the range of α is infinite, it may be that the principal set of equipossible primary alternatives under α with respect to β is finite. In that case $P(\alpha, \beta)$ is the ratio of the number of alternatives in this set which entail β to the total number of alternatives in it. (b) It may be that the principal set of equipossible primary alternatives under α with respect to β is itself infinite, *e.g.*, they may involve the different values of a *continuous* variable. Mr. Kneale says that, in such cases, the measure of a range has to be conceived as the measure of 'a region in a configuration-space', *i.e.*, by analogy with the length of a line or the area of a surface or the volume of a solid, and $P(\alpha, \beta)$ has to be regarded as the ratio between the measures of two such regions.

Mr. Kneale does not give us much help in connexion with the *general* notion of a configuration-space in probability or with the question how regions in it are supposed to be measured. He does discuss very elaborately certain well-known paradoxes of 'geometrical' probability. His discussion of Bertrand's Paradox about the probability of a chord 'drawn at random in a circle' being longer than the side of the inscribed equilateral triangle seems to me very illuminating.

Reverting to the general topic of the Range Theory, we must note that Mr. Kneale is perfectly well aware that no-one can produce an example of a principal set of equipossible primary alternatives falling under any natural characteristic, such as *human*. He is claiming only to *analyze the meaning* of '$P(\alpha, \beta) = p$'. He does not imagine that a knowledge of this will enable one to *determine the value* of $P(\alpha, \beta)$ *a priori* when α, *e.g.*, stands for *human*, and β, *e.g.*, stands for *male*. All probability-rules about open classes resemble laws of nature, in that they can be inferred only by ampliative induction. The Frequentists are quite right in saying that the evidence for such rules is observed frequencies. Their mistake is to hold that what is inferred is definable in terms of frequency. This mistake is analogous to that of thinking that a *law* is a 100 per cent *de facto* association. The assumption at the back of both mistakes is that the conclusion of an inference must be a proposition of the same type as the premisses.

A CRITICAL NOTICE OF 'PROBABILITY AND INDUCTION'

If. Mr. Kneale is right, the conclusions of all ampliative inductions are different in kind from their premisses. For the premisses are in all cases about *matters of fact*: whilst the conclusions, according to him, are *principles of modality*, whether they be laws or probability-rules.

The next important question discussed by Mr. Kneale is whether it can be shown, by means of the calculus of probability, that ampliative induction leads in favourable cases to conclusions which are highly probable in the sense contemplated by that calculus. After examining the so-called 'inversion' of Bernoulli's Theorem, Laplace's Rules of Succession, and Keynes's Principle of Limited Variety, with unfavourable results, Mr. Kneale brings forward what he considers to be a fundamental objection to all attempts to justify ampliative induction within the theory of chances.

His argument may be put as follows. The propositions which we try to establish by ampliative induction are either laws or probability-rules. Let us begin with the laws. Suppose that the law to be established is All S is P. We have observed n instances of S, say $S_1, S_2, \ldots S_n$, and have found that all of them are P. It is claimed that we can show by using Bayes's Theorem that the probability that All S is P, given the conjunctive proposition S_1 is P-and-S_2 is P-and...S_n is P, approaches to 1 as n is indefinitely increased. Now it is admitted that the argument requires the fulfilment of the following two conditions. (i) That the *antecedent* probability of All S is P is greater than some number which is itself greater than 0. (ii) That the probability of the conjunctive proposition, given that the law is *false*, approaches indefinitely to 0 as n is indefinitely increased. It is argued that the second condition is fulfilled because the probability of this conjunctive proposition, on this hypothesis, is the product of n terms, each of which is a proper fraction in a sequence whose successive terms do not tend to unity as n is indefinitely increased. Now suppose, if possible, that the law All S is P were just an endless factual conjunction of singular propositions *i.e.*, that it was the proposition S_1 is P-and-S_2 is P-and...S_n is P-and.... Then by precisely the same argument which proves that the second condition is fulfilled we could prove that the first condition is *not*. On this interpretation of law the antecedent probability of any law would be 0. Therefore, unless the argument is to break down at the first move, it must assume (what Mr. Kneale claims to have shown independently) that laws are *not* endless conjunctions of singular propositions. This is the first step in Mr. Kneale's argument.

The next step is this. The only acceptable alternative analysis of laws is that they are modal principles of necessary connexion between attributes. But it is meaningless to assign a probability, in the sense in which that term is used in the theory of chances, to a modal principle. Probability, in that sense, presupposes real objective alternative possibilities; and it is plainly meaningless to regard a principle of necessary connexion as one alternative possibility among others. Therefore a law has no antecedent probability (and of course no consequent probability) in the sense required by the above attempt to apply Bayes's Theorem.

Now, on Mr. Kneale's view, probability-rules are also modal principles concerning the possibilities that are left open by laws. Therefore, they too can have no probability in the sense required in the theory of chances; and therefore there can be no question of showing that the process of ampliative induction from observed frequencies confers upon probability-rules a high probability in that sense.

The last topic which Mr. Kneale discusses in this Part is the theory of sampling from finite populations. Here the conclusion that the population as a whole contains a certain proportion of instances of a given characteristic has a probability in the sense required for the application of inverse-probability arguments. But in practice such arguments are seldom applicable, since we do not generally know the antecedent probabilities of the various alternative possible proportions.

4. *Probability of inductive science*

The question discussed in this Part may be put as follows. Is there any sense of 'justification' in which ampliative induction *needs* justification? If so, *can* it be justified in that sense? And, if so, *how* can it be justifed? The discussion is inevitably somewhat complicated. For, in the first place, we have to deal with (1) *primary*, and (2) *secondary* inductions, *i.e.*, those which directly induce laws or probability-rules from observations, and those which establish explanatory theories on the basis of such laws. Then, within the discussion of primary induction, we have to consider the establishment of (1.1) *laws*, and (1.2) *probability-rules*. Moreover, a law may be either (1.11) of the purely *qualitative* form All S is P, or (1.12) of the *functional* form $Y = f(X)$. Lastly, the results of an inductive argument, whether primary or secondary, are not just rationally acceptable or unacceptable. According to circumstances they may be *more or less* rationally acceptable.

The ground has already been cleared to the following extent. We know that it is absurd to think that ampliative induction can be justified in the sense that its conclusions can be *deduced demonstratively* from its premisses. We also know that it is absurd to think that it can be justified in the sense that its conclusions can be shown to have a *high probability* (as understood by the theory of chances) in relation to its premisses. Some persons have concluded from this that the question: 'Is ampliative induction justifiable, and if so, how?' is a meaningless question, which would cease to be asked if these negative facts were pointed out and appreciated. Mr. Kneale does not accept this conclusion. According to him, induction is a 'policy' which one might or might not adopt in certain situations in which we are all very often placed. The question is whether we can show, apart from all reference to the truth or the probability (in the technical sense) of inductive conclusions, that inductive policy is the one which a sensible person, aware of his own needs, resources, and limitations, 'could not fail to choose'. I think that the phrase in inverted commas is highly ambiguous, and I am not perfectly sure what Mr. Kneale means by it. The meaning may become clearer to the reader when he has seen the application.

What then *is* the policy of primary induction (a) in regard to laws of the form All S is P, (b) in regard to laws of the form $Y = f(X)$, and (c) in regard to probability-rules?

(a) Let us use the symbol 'S_0' to denote *observed* instances of S, and similarly *mutatis mutandis* for 'P_0' and 'Q_0'. Suppose that the empirical facts can be stated as follows. All S_0 is P. All S_0 is Q. Some P_0 is neither S nor Q. Some Q_0 is neither S nor P. The only laws which are compatible with these observations are All S is P, All SQ is P, All S is Q, and All SP is Q. The most timid policy would be to formulate no laws at all. Still playing for safety, one might formulate the laws All SQ is P and All SP is Q. The boldest policy consistent with the observations would be to accept tentatively the laws All S is P and All S is Q. The other aspect of the policy would be to look out for instances of S which were not P and instances of S which were not Q. But, unless and until such instances were found, it would be contrary to the policy to be content with the more restricted laws All SQ is P and All SP is Q. The policy here may be summed up as follows. (i) Act in all relevant circumstances on the assumption that combinations of characteristics of which you have found no instances in spite of seeking for them are incompatible. But (ii) continue to look for instances of such conjunctions, and be prepared to admit extensions of the range of what you have hitherto taken to be possible so far and only so far as fresh observations compel you to do so.

(b) The inductive policy in the case of functional laws is as follows. Act on the assumption that the law connecting the values of Y with the associated values of X is the 'simplest' consistent with the observations made up to date, but be on the look-out for new pairs of associated values which this curve fails to fit. Here one curve is 'simpler' than another if it requires fewer independent parameters to determine it completely; in this sense a straight line is simpler than a circle, a circle than a parabola, and a parabola than an ellipse or an hyperbola.

(c) In the case of probability-rules the inductive policy is as follows. If the relative frequency of instances of α which are β among all the instances of α which have been observed is p, act on the assumption that the value of $P(\alpha, \beta)$ is p. What we are trying to do in such cases, on Mr. Kneale's interpretation of $P(\alpha, \beta)$, is to make the best guess that we can, on the basis of the available statistical evidence, as to the ratio of the range of possibilities under $\alpha\beta$, left open by all the principles of necessitation and exclusion, to the range of possibilities under α, left open by those principles. It should be noted that to act on this policy is equivalent to assuming that value of $P(\alpha, \beta)$ which gives the maximum probability to the actual frequency of β's found in the finite class of n observed instances of α, i.e., which maximises the value of $P\left(^{\alpha}_{c}\sigma_n, {}^{\beta}\rho_p\right)$.

The policy in all three cases falls under the following general maxim. In any case where you have to act, either practically or theoretically, on partial knowledge, act as if you knew that the boundaries of possibility lie as nearly as may be to the actual associations and dissociations and proportions which you have observed and critically tested up to date.

Why, and in what sense, is this policy 'reasonable' or 'justifiable'? We are often in a position where our practical or theoretical interests oblige us to treat an object, of which we *know* only that it is or will be an instance of α, as if it were or would be β or as if it were or would be non-β. The only way in which we can do this is by assuming the truth of a relevant law or probability-rule on the basis of our observations up to date. If all the observed instances of α have been β, it is for various reasons more profitable to assume the law that All α's are β than to assume any less sweeping law, such as All $\alpha\gamma$'s are β, or to assume merely that a certain percentage of α's are β. The advantages are the following. If the supposition should be false, it is likely to be sooner refuted by counter-instances than any of the less sweeping suppositions compatible with the at present known facts. If, on the other hand, it should be true, it will be more powerful as a premiss for inference than any of these less sweeping assumptions. To this it may be added that, if one were to postulate anything but the strongest law consistent with the known facts, it is difficult to see where one could reasonably draw a line, since any set of observed instances of S which were all P would have innumerable properties in common beside S and P.

The justification is very similar in the case of functional laws. Suppose, e.g., that you have observed n pairs of associated values of Y and X, and have found that they all fall on a certain straight line $y = a_0 + a_1 x$. The law connecting Y with X must be represented either by this straight line or by one of the innumerable curves of higher order which cut it in at least those n points but diverge from it elsewhere. If the linear hypothesis should be false, a single unfavourable further observation will suffice definitely to refute it; but, however, the $n+1$th observation may turn out, it will be consistent with innumerable more complicated laws, between which one would have no reasonable ground for choosing.

I doubt whether I fully understand Mr. Kneale's argument to justify the procedure of assigning to $P(\alpha, \beta)$ the value p, when one has examined n instances of α and found that they contain a proportion p of β's. It certainly starts from the proposition (which is easily proved) that to assign any *other* value than p to $P(\alpha, \beta)$ would entail a *lower* value for the probability that a set of n instances of α would contain the observed proportion p of β's. The argument then seems to run as follows. By definition, the latter probability is the ratio of the range of possibilities under the property of being an n-fold set of α's containing a proportion p of β's to the range of alternatives under the property of being an n-fold set of α's containing *any* proportion of β's from 0 to 1. Now, it is alleged, the extent of the former range is *independent* of the value of $P(\alpha, \beta)$, whilst the extent of the latter range is *dependent* on the value of $P(\alpha, \beta)$. It follows that the value of $P(\alpha, \beta)$ which makes this ratio a maximum is the value which makes its denominator a *minimum*. Therefore, to assign as the value of P(α, β) the observed frequency p, with which instances of β have occurred in the n-fold set of α's examined, is equivalent to assuming that the range of possibilities under the property of being an n-fold set of α's containing *any* proportion of β's is as *narrow* as is consistent with the observations.

The step in this argument which I do not understand is the statement that the range of alternatives under the property of being an n-fold set of α's containing a proportion p of β's is *independent* of the value of P(α, β), whilst the range of alternatives under the property of being an n-fold set of α's containing *any* proportion of β's from 0 to 1 is *dependent* on the value of P(α, β). Let us take, *e.g.*, a finite class of N α's, and suppose it contains exactly $Nq\beta$'s. Then the value of P(α, β) is q. Now the number of possible n-fold subclasses containing a proportion p of β's would seem to be

$$^{Nq}C_{np} \, ^{N(1-q)}C_{n(1-p)},$$

i.e., to be *dependent* on q, the value of P(α, β). And the number of possible n-fold sub-classes of *any* possible constitution in respect of β would seem to be $^{N}C_{n}$, *i.e.*, to be *independent* of q. This is the exact opposite of Mr. Kneale's statement. I suppose that there must be a simple misunderstanding somewhere, but I cannot make out where it lies.

The last topic to be discussed under this head is the varying degrees of irrationality which are involved in departing from the inductive policy under various circumstances. Here Mr. Kneale distinguishes two defects in a hypothesis, which he calls 'Extravagance' and 'Negligence'. The former applies both to assumptions of law and assumptions of probability-rules. The latter applies only to the case of laws. I will take them in turn.

As we have seen, if we follow the inductive policy we are in effect ascribing to P(α, β) that value which maximizes the probability that an n-fold set of α's would have the proportion of β's which it has in fact been found to have. Mr. Kneale defines the 'extravagance' of any departure from the inductive policy as the ratio of the diminution of this probability, entailed by that departure, to the maximal value, which it has if the policy is followed exactly. It is easy to show that, with this definition, the extravagance of a given departure from that value of P(α, β) which the inductive policy would dictate increases with the size of the sample observed. The formula covers the two extreme cases of 100 per cent and

0 per cent observed frequencies of β among α's, where the inductive policy would be to postulate a *law*.

'Negligence', in the technical sense, consists in assuming only a probability-rule where the observations are consistent with a law; or in assuming a law with a more restricted subject or a less determinate predicate when the observations are compatible with a law with a less restricted subject or a more determinate predicate.

So much for Mr. Kneale's views on the 'justification' or primary induction; it remains to consider the 'justification' of secondary induction.

A theory is put forward to *explain* laws and probability-rules which have been or may be established by primary induction. A successful theory introduces *simplification* in two different, though connected, senses. In the first place, it must, of course, entail all the primary generalizations which it is put forward to explain, and others too which can be tested. Now it seems clear that the question whether a generalization, which is entailed by a theory, was established by primary induction *before* or *after* the putting forward of that theory cannot be of any logical relevance to the support which it gives to the theory. If a newly drawn consequence is to support the theory, it must be verified by primary induction before it can do so; and, when once this has been done, it is in the same position as the already verified generalizations which the theory was originally put forward to explain. Mr. Kneale concludes that a theory is not worth serious consideration unless it entails an unlimited number of testable consequences. If this be granted, the first sense in which a successful theory simplifies is that it restricts the realm of possibility more than is done by any finite number of empirical generalizations entailed by it.

The second sense in which a successful theory simplifies is that it reduces the number of independent concepts, and thus reduces the number of independent propositions, which we have to accept. An example is the unification of electricity, magnetism, light, etc., by Maxwell's Theory.

If the acceptability of a theory is to rest on its having been formulated and tested in accordance with a policy indispensable to pursuing an end which we seek, we must ask what that end is. Now theories certainly have the following two uses. A theory suggests subjects which it may be profitable to investigate by primary induction, and thus has an important directive use. Again, when it is shown that a number of primary generalizations are all consequences of a theory, the special evidence for each is reinforced by the evidence for all the rest. But, Mr. Kneale holds, these two valuable services which theories render are not the ultimate motive for theorizing by scientists. Men desire explanation for its own sake, and this desire is the main motive with pure scientists. The satisfaction derived from a good theory is in certain ways analogous to aesthetic satisfaction. But there are important differences. Scientific theorizing is not *free* construction, like musical composition. The scientist wants his theories to be *true*, and the minimum condition is that they shall be consistent with all known empirical facts. Moreover, he has the ideal of a single *all-embracing* theory, under which all possible empirical generalizations can be subsumed, and to which there is no alternative. Why men should have this ideal we do not know, but it is a fact that great scientists do have it. Secondary induction is justified in so far as it is the only policy by which we can set about realizing this ideal. We have no guarantee that it is realizable, and, if we happened to have realized it, we could never *know* that we had done so. But, if there *is* a single system of natural necessity, then the procedure of secondary induction is the only policy by which we can hope to approximate our beliefs to it.

II. CERTAIN CHARACTERISTIC DOCTRINES OF MR. KNEALE

As we have seen, Mr. Kneale holds the following unfashionable views. (i) That laws of nature are principles of necessity, of the same nature as the proposition: A surface cannot be at the same time red and green all over; though, unlike that proposition, they are incapable of being revealed by intuitive induction and known *a priori*. (ii) That such propositions are not merely linguistic. It will be convenient to consider his views on these two points in the opposite order to that in which I have stated them.

1. *Principles of modality are not merely linguistic*

Principles are truths about the possibility or impossibility of certain characteristics being combined in facts of a certain structure. They are more fundamental than facts, in the sense that it depends on them what are possible facts and what are not. On the other hand, we could not formulate any principle unless we were acquainted with, and had formulated, some facts. For, in the first place, we could not be aware of any characteristic unless we were acquainted with facts in which it is a component. And, secondly, unless we had formulated some facts, we should have no means of symbolizing the structure of various kinds of possible fact. Mr. Kneale holds that all knowledge of singular negative facts, *e.g.*, the fact that the paper on which I am writing is not blue, involves knowing principles as well as facts. I must know, *e.g.*, the fact that this paper is white. But I must also know that it is possible for paper to be blue, and that being white all over is incompatible with being blue all over. This seems to me to be obviously true.

Consider now the allegation that the sentence 'It is impossible for anything to be at once red and green all over' merely records a linguistic convention that no sentence of the form 'X is at once red and green all over' is to be used. Certainly it is a matter of linguistic convention that 'red' means what it does in English and that 'green' means what it does in English. It is quite possible, *e.g.*, that 'red' should have meant what it does now mean, and that 'green' should have meant what is now meant by 'scarlet' or what is now meant by 'hot'. In that case the sentence 'X is at once red and green all over' would have been permissible. The fact that it *is* not permissible depends on the fact that 'red' and 'green' at present mean two characteristics which are in themselves incompatible spatio-temporally. And it *would have been* permissible only if the meaning of one or of both of these words had been such that they name characteristics which are in themselves spatio-temporally compatible. *Any* language which contains names for the characteristics of which the words 'red' and 'green' are the names in contemporary English will have to use those words in accordance with a rule corresponding to the English rule about the use of 'red' and 'green'. And that is because the rule states a principle concerning the characteristics of which these words are names. This, again, seems to me to be quite obviously true.

Mr. Kneale adds the following argument, which I give for what it may be worth. When one learns how to use a word, *e.g.*, 'red', correctly, an essential part of what one learns is *not* to use it *unless* a certain condition C is fulfilled. In order to act on this knowledge one must be able to recognize cases in which C is *not* fulfilled. But one can never know a negative singular fact without using one's knowledge of a principle of incompatibility. Therefore ability to avoid using a word incorrectly involves knowing principles of modality.

2. Laws are principles of modality

Mr. Kneale's view of the nature of laws may be compared with democracy in at least one respect. There are strong *prima facie* objections to it, and the only good arguments for it are the arguments against all the alternatives. Accordingly, we shall be concerned mainly with his criticisms of alternative analyses of law, and with his attempt to answer the *prima facie* objections to his own analysis of it.

The two alternative analyses which are worth serious consideration are the following. (i) It might be alleged that the law: All S is P can be identified with the unrestricted factual proposition: Every instance of S that has been, is, or will be, has been P or is P or will be P, as the case may be. (ii) It has been alleged that laws, though expressed by sentences in the indicative, like 'All S is P', are not really propositions at all. They are prescriptions, which would be less misleadingly expressed by a sentence in the imperative, *e.g.*, 'Whenever you meet with an instance of S and do not know whether it is P or not, act on the assumption that it is P'.

It has been objected to the purely factual analysis of law that, if it were true, no law could conceivably be verified by experience, and that this would entail that all nomic, sentences are meaningless. Mr. Kneale does not accept this argument, because he rejects this criterion of significance. He points out that the statement 'There is at least one instance of S which is not P' is certainly capable in principle of being verified, and is therefore significant by this criterion. It would be strange if this significant statement should have no significant contradictory.

Mr. Kneale's own objection is radical. Laws are not facts at all, and therefore not facts of the form alleged. To state a law properly we need a *conditional* sentence, not a mere sentence in the indicative. If it is a *law* that all S is P, then anything that *had* been, or that *might* now be, or that *should* in future be an instance of S *would* have been P, or *would* now be P, or *would* then be P.

Since nomic sentences are not statements of fact, anyone who denies that they are statements of modal principles of necessity, is practically forced to hold that they are not really statements at all but are disguised prescriptions. Now a prescription is either a command or an admonition. If Boyle's Law, *e.g.*, is a command, like 'form fours', one would wish to know, before obeying it, who issues the command, what authority he has

A CRITICAL NOTICE OF 'PROBABILITY AND INDUCTION'

for doing so, and what penalties he can and will inflict in case of disobedience. Obviously there is no answer to these legitimate questions in the case of a law of nature. If on the other hand, it is an admonition, like 'Cast not a clout till May be out', it is reasonable to ask what advantages are to be derived or what disadvantages are to be avoided by following the advice. If the person who gives us this advice answers that acting in this way will enable one to make successful predictions, he appears to be enunciating a law of nature in a non-prescriptive sense. If he answers that this is the policy which scientists do pursue, one can raise the following two supplementary questions. 'Do you mean merely to put on record the way in which scientists have in fact behaved up to date, or are you enunciating a *law*, in the non-prescriptive sense, about the behaviour of a certain class of human beings?' And whichever answer is given to this question, one can then ask: 'What is the relevance of your answer to the question why should I follow your advice in this matter?' To put it shortly, is there any reasonable ground for following the advice to act as if S were P whenever you meet an instance of S except that there is reason to believe that: All S is P is a law of nature in the non-prescriptive sense?

Finally, we can consider Mr. Kneale's answer to the *prima facie* objection that laws of nature cannot be principles of necessity, because any principle of necessity would be capable of being known a *priori* whilst no law of nature can be so known.

The objection is often put in the form that, if you can imagine an instance of S which is not P, then S cannot necessitate P. I shall state what I believe to be Mr. Kneale's main contentions in my own way and with my own examples.

In the first place, an example from pure mathematics has a certain relevance to the objection. Take the proposition that the square-root of 2 is irrational. This means that there are no two integers m and n, such that the ratio of m to n (reduced to its lowest terms) squared is equal to 2. Now this proposition is true and necessary and easily proved. But there is an important sense in which it is perfectly easy to 'imagine what it would be like' if the proposition were false. One can imagine oneself applying to the number 2 the ordinary process for extracting a square-root, and finding that it came to an end after a finite number of steps, as it does, *e.g.*, after two steps if applied to the number 841. This example is useful as a counter-instance to the general principle that a proposition cannot be necessary if one can 'imagine what it would be like' for it to be false. But it would be a mistake to rest any positive analogy on it; for laws are certainly different in kind from propositions about numbers, even if they be of the same kind as propositions which can be established by intuitive induction.

Coming to Mr. Kneale's main contention, I find it easier to give an account of the explicit premisses, the main steps of the argument, and the conclusion, than to indicate the precise connexion between them. Mr. Kneale begins by pointing out that natural laws are concerned with *perceptual* events and things, *e.g.*, flashes of lightning or samples of ammonia, and not with merely *sensible* events and objects, such as the visual sense-datum which is presented to a person when he sees a flash of lightning or the olfactory sense-datum which is presented to him when he smells a whiff of ammonia.

He then considers the relation between sensation and sense perception. He accepts the conclusion that the statement 'X is seeing the perceptual-object O' implies (i) that X is sensing a certain visual sense-datum, and (ii) that this, in some sense, 'belongs' to a certain physical object which can be correctly described as, or named by 'O'. In considering what meaning to attach to the word 'belongs' in this context he rejects any view which would imply that it is intelligible to suggest that there might be a sense-datum which was not sensed by anyone. After considering and rejecting various alternative theories, Mr. Kneale says that he thinks that the following is 'correct so far as it goes'. Statements involving names and descriptions of perceptual objects and their properties are not *reducible to* statements about actual and possible sensations; but they are an *appropriate device* for referring briefly and compendiously to innumerable propositions about the sensations which would be experienced under innumerable different conditions. It must be noted, however, that an unlimited number of these propositions about sensations would be of the form: 'If a person *had* been in such and such a place at such and such a time and *had* then and there done such and such things, he *would* have had such and such sensations', where no-one in fact was there or did those things at that time. Such propositions about the consequences of unfulfilled conditions seem to involve, either directly or at a later move, propositions to the effect that one kind of sensible experience would *necessitate* a sensible experience of a certain other kind.

Mr. Kneale concludes from all this (what is undoubtedly true) that perceptual-object words, like 'lightning', 'ammonia', 'flexible', 'soluble in water', and so on, obey utterly different rules from words and phrases about individual sense-data and their qualities. He suggests that the opinion that laws of nature would be knowable *a priori* if they were principles of necessity has arisen only because people have either failed to notice that such laws are concerned with perceptual objects and their properties, or have failed to see that propositions about the latter differ fundamentally from propositions about sense-data and their qualities.

Now it is this last vitally important contention which seems to me not to have been adequately developed and illustrated by Mr. Kneale. I think that he ought to have done the following three things. (i) To produce evidence that competent contemporary philosophers who disagree with his views on the nature of laws *do in fact* fail to see the distinction in question. For my part, I very much doubt that they do. (ii) To show us *why* principles of necessary connexion concerning sense-data and their qualities might be expected to be capable of being known *a priori*. And (iii) to indicate *how precisely* the admitted differences between sense-data and their qualities, on the one hand, and perceptual objects and their properties, on the other, make it impossible that any principle of necessary connexion concerning the latter should be known *a priori*. If Mr. Kneale has given an adequate answer to questions (ii) and (iii), I must confess that I do not understand it clearly enough to be able to convey it to the reader.

I greatly hope that Mr. Kneale will enlighten us further on these points. In the meanwhile he may be heartily congratulated and thanked for the bold, original, and extremely well-written contribution which he has made to one of the hardest and weightiest of the problems of philosophy.

GEORG HENRIK VON WRIGHT

BROAD ON INDUCTION AND PROBABILITY

If I had to name the most important contributions to inductive logic from the period between the two great wars, I should without hesitation mention the following ones: Keynes' *A Treatise on Probability* (1921), Nicod's *Le Problème Logique de l'Induction* (1923), the chapters on induction and causality in the second and third volumes of Johnson's *Logic* (1922, 1924), and Broad's papers, 'Induction and Probability' (I–II, 1918, 1920), 'The Principles of Problematic Induction' (1927), and 'The Principles of Demonstrative Induction' (I–II, 1930). To these might be added F.P. Ramsey's posthumous essay 'Truth and Probability' and R.A. Fisher's criticism, in various publications, of the Bayes-Laplacean doctrine of so-called inverse probability. I believe that few informed people would disagree with the choice.

It has struck me that all the authors mentioned are Cambridge men. (Nicod studied and worked at Cambridge during the first great war, and his published work is in the spirit of British philosophy.) There thus exists something which can be called a Cambridge tradition in modern inductive logic.

If Broad's writings on induction have remained less known than some of his other contributions to philosophy and less influential than the works of some of the other authors mentioned above, *one* reason for this is that Broad never has published a book on the subject. It is very much to be hoped that, for the benefit of future students, Broad's chief papers on induction and probability will be collected in a single volume – possibly with some contributions of his to other branches of logical theory.

The following is a list of Broad's writings on inductive logic:

[1] 'The Relation between Induction and Probability I–II', *Mind* **27** (1918) 389–404 and **29** (1920) 11–45 (present volume, pp. 1–52).

[2] Critical notice on J. M. Keynes, *A Treatise on Probability*, *Mind* **31** (1922) 72–85 (present volume, pp. 53–68).

[3] Critical notice on W.E. Johnson, *Logic, Part II*, *Mind* **31** (1922) 496–510 (present volume, pp. 69–72).
[4] 'Mr. Johnson on the Logical Foundations of Science I–II', *Mind* **33** (1924) 242–261 and 369–384 (present volume pp. 73–85).
[5] *The Philosophy of Francis Bacon*, Cambridge 1926. Reprinted in *Ethics and the History of Philosophy*, 1952.
[6] 'The Principles of Problematic Induction', *Proceedings of the Aristotelian Society* **28** (1927–1928) 1–46 (present volume, pp. 86–126).
[7] 'The Principles of Demonstrative Induction I–II', *Mind* **39** (1930) 302–317 and 426–439 (present volume, pp. 127–158).
[8] 'Mechanical and Teleological Causation', *Proceedings of the Aristotelian Society*, suppl. vol. **14** (1935) 83–112 (present volume, pp. 159–183). (Part of a symposium with this title, with C.A. Mace, G.F. Stout, and A.C. Ewing as the other participants.)
[9] Critical notice on R. von Mises, *Wahrscheinlichkeit, Statistik und Wahrheit*, *Mind* **46** (1937) 478–491 (present volume, pp. 184–200).
[10] 'Hr. Von Wright on the Logic of Induction I–III', *Mind* **53** (1944) pp. 1–24, pp. 97–119, and pp. 193–214.
[11] Critical notice on W. Kneale, *Probability and Induction*, *Mind* **59** (1950) 94–115 (present volume, pp. 201–227).

In the following I shall refer to these works using the numbers within brackets [ ]. I shall use [1^I] and [1^{II}] to refer to part one and part two respectively of [1]; and similarly [7^I] and [7^{II}].

As an inductive logician, Broad has acknowledged[1] his indebtedness both to Keynes and to Johnson. A few words about the relation of his work to theirs may be called for.

Both parts of [1] were published before the appearance of Keynes' *Treatise*. There are two important points, on which Broad can be said to have anticipated Keynes' work. One is the mathematical treatment of the probability-relation between a generalization and its confirming instances. By virtue of it, Broad ought to be regarded as the founder of that branch of modern inductive logic, which may appropriately be termed Confirmation-Theory.[2] The second point is the idea of Limited Independent Variety.[3] This idea, in inductive logic, ultimately goes back to Bacon. With Broad in [1] it takes the form of a theory of natural kinds, with Keynes that of a theory of generic properties or of 'legal atoms' in nature. The two theories are, as far as I can see, rather far removed from each other. In [6] Broad gives some interesting further developments of the *Keynesian* form of the idea of limited variety. (See below under I and II.)

The relation of Broad's work in inductive logic to Johnson's is of a more complex kind. As far as technical terminology and fundamental distinctions are concerned Broad obviously owes much to Johnson. His paper [7] very closely follows Johnson's lines, and it seems a fair characterization to say that its primary aim is to improve upon Johnson's not too clearly formulated theory of demonstrative induction. It accomplishes, however, much more than this. The first part of [7] opens up a new and fruitful approach to the old subject of formalizing the 'canons' or 'methods' of induction by elimination. The second part is less original; I cannot suppress a feeling that the subject would have profited more, had Broad attempted to give an independent treatment of the theory of inductively correlated variations, rather than a perfected statement of the Johnsonian 'figures' which, as Broad himself shows, rest on the use of untenable postulates. (See below under III.)

The rest of the present essay is divided into four sections. In the first three sections I examine, in order, Broad's three main publications in the field of inductive logic. Those are the three papers mentioned in the opening paragraph above and subsequently listed under numbers [1], [6], and [7]. Having examined them, I draw in the concluding section attention to some points in the remaining eight relevant publications which seem to me particularly interesting. All through my essay I have tried to concentrate on questions which I consider important from the point of view of the subject as a whole.

I

In [1] Broad proposes[4]

> to prove three points, which, if they can be established, are of great importance to the logic of inductive inference. They are (1) that *unless* inductive conclusions be expressed in terms of probability all inductive inference involves a formal fallacy; (2) that the degree of belief which we actually attach to the conclusions of well-established inductions cannot be justified by any known principle of probability, unless some further premise about the physical world be assumed;

and (3) that it is extremely difficult to state this premise so that it shall be at once plausible and non-tautologous.

I shall here deal with these three points in order.

1. It is not quite clear to me what Broad means by the 'formal fallacy' which an inductive argument is alleged to commit, unless its conclusion be stated 'in terms of probability'. But, as far as I can see, the 'formal fallacy' is nothing but the fact that (ampliative) induction is an *inconclusive* type of argument. It would seem to follow from this, that an inductive argument which does *not* commit a fallacy must be *conclusive*.

What then is the correct way of formulating an inductive argument? Many logicians have thought that by introducing an additional premiss – often referred to under the name of the Principle of the Uniformity of Nature or Law of Causation – the formal fallacy of ordinary inductive reasoning can be avoided. This, however, is disputed[5] by Broad. The argument which he gives[6] I find neither clear nor convincing. I should have thought that, unless the form or content of the additional premiss is restricted in advance, it is perfectly possible to state inductions in the form of demonstrations without mentioning 'probability' in the conclusion. This, I believe, is implicitly admitted by Broad too in [7].

Broad seems to think that the formulation of inductive conclusions in terms of probability 'accords with what we actually believe when we reflect'.[7] Now, what I think we all admit upon reflection is that, in so-called ampliative or problematic inductions, the conclusions do not follow logically from their premisses, *i.e.* that the falsehood of the conclusion is consistent with the truth of the premisses. But to clothe this admission in a probabilistic terminology – as is often done by philosophers – is not an altogether good idea. It leads to conflict with a view on *certainty* which seems to me quite sound to entertain. That I shall die is *certain* ('as certain as anything') although my immortality is logically compatible with the mortal nature of any number of men in the past. There are numerous inductions, the conclusions of which are in this same sense 'certain'. To say that they are, after all, 'probable' only, is but another way of saying that they are not *logically* 'certain'. But this is a bad way of putting it. It tends to obscure an important distinction, *viz.* the distinction between those inductive conclusions which are *certain*, though not logically certain, and those which are not certain in *any* sense of the word and which for *that* reason may be called *probable* only.

It seems to me therefore that the formulation of inductive conclusions in terms of probability is not only not necessary from the point of view of logic, but actually conflicts with a perfectly respectable use of 'certainty' in connexion with induction. I believe that this criticism is of some relevance to a discussion of the so-called justification of induction. It does not, however, affect the rest of Broad's argumentation in [1] about the relation of induction to probability.

2. Let us, for present purposes, by the *factual premiss* of an inductive argument understand a proposition of the form 'All observed S's have been P'. Let the conclusion, stated in terms of probability, be either of the form (i) 'It is probable to degree p that all S's whatever will be P' or of the form (ii) 'It is probable to degree p that the next S to be observed will be P'.

I believe that the problem behind the second point which Broad wishes to establish in [1] can be formulated as follows:

If an inductive argument is to be formally in order, *i.e.* conclusive, it is *necessary* that to its factual premiss be added some further premiss or premisses which are formal principles of probability. These principles are a sort of logical laws; they hold true 'in all possible worlds'.[8] The question now is, whether the addition of such formal principles to the factual premiss also is *sufficient* to make the argument conclusive. Broad shows that the answer is negative. In order to make inductive arguments conclusive still further premisses are needed. These premisses are not logical laws but material assumptions concerning the constitution of nature. In showing all this Broad is, I think, perfectly successful.

Broad distinguishes between induction 'by simple enumeration' and induction 'by the hypothetical method',[9] and proceeds to consider each case separately. The distinction, though in itself important, appears to me in this connexion somewhat confusing. The essential difference between the two cases which Broad examines is that they represent two different types of approach to the problem of how to assign a probability to an inductive conclusion. The first approach is the classical doctrine of so-called Inverse Probability, founded by Bayes and Laplace. The second is a new approach of which Broad and Keynes are the pioneers.

BROAD ON INDUCTION AND PROBABILITY

A. *Induction by simple enumeration*

Let the factual premiss of the inductive argument be that m counters have been drawn, without replacement, from a bag and all of them found to be white. What is the probability that the next counter which will be drawn is white? And what is the probability that all the n counters, which originally were in the bag, are white?

According to the Laplacean doctrine, the answer to the first question is given by the formula $\frac{m+1}{m+2}$ and the answer to the second question by the formula $\frac{m+1}{n+1}$.

Broad states the formal and material assumptions which he thinks are needed for a correct derivation of the two formulae. Among the assumptions there are two crucial ones about *equi-probability*.

The first assumption of equi-probability is to the effect that all the possible proportions of white counters in the bag are initially equally probable. Of this assumption Broad in [1] takes the view that it is true on logical grounds alone. Or more precisely: he views it as a legitimate application of a 'Principle of Indifference' which he takes to be logically necessary.[10] He does not note the difficulty caused by the fact that several different so-called *constitutions* of the bag may answer to one and the same *proportion* of white counters in it. In [6] Broad is aware of this difficulty. (See below p. 244–245.)

The second assumption of equi-probability is to the effect that each individual counter in the bag has the same probability of being drawn.[11] This assumption is 'material', *i.e.* its truth does not follow from laws of logic and necessary principles of probability alone. But is it really needed for a correct derivation of Laplace's formulae? As far as I can see it is *not* (unconditionally) needed. It is needed only if we wish to make the transition from the proposition that the *proportion* of white counters in the bag is p to the proposition that the *probability* of drawing a white counter from the bag is p. It is needed, in other words, to make the definition of a degree of probability as a ratio among favourable and unfavourable alternatives of equal probability applicable to the case under discussion. But the problem can be treated independently of this way of defining degrees of probability, within a so-called uninter-

233

preted calculus. Then Broad's two assumptions about equi-probability may be replaced by the *one* assumption that any given value of the probability of drawing a white counter is initially as probable as any other value. This assumption, in my opinion at least, is 'material'.

Broad now goes on to consider[12] 'how far the attempt to establish laws of nature by simple enumeration is parallel to the artificial example just dealt with' of the counters in the bag. Two cases are distinguished[13]: laws about the qualities of classes of substances ('All crows are black') and laws about the connexion of events ('All rises of temperature are followed by expansion').

One main difference between the artificial and the 'natural' cases has to do with the second assumption of equi-probability mentioned above. All counters in the bag *may* have the same probability of being drawn, but all individual crows certainly have not the same probability of being observed. This is so for two reasons. First, because 'we clearly cannot have observed any of the crows that begin to exist after the moment when we make the last observation which we take into account when we make our induction'.[14] And secondly, because our observations are confined to a restricted region in space, and 'the blackness of the observed crows may not be an attribute of all crows but may be true only of crows in a certain area'.[15] The same difference between the artificial and the 'natural' cases is there for laws about events. Broad also points out other differences, but I shall not mention them here.

The second assumption of equi-probability, it was just said, is needed only if we wish to associate with the formulae $\frac{m+1}{m+2}$ and $\frac{m+1}{n+1}$ the 'classical' definition of probability as a ratio among equi-possible unit-alternatives. In the case of crows and blackness this would amount to saying that the proposition that there is a probability to degree p that a random crow is black *means* that any one crow is as likely as any other to be observed and a proportion p of them all are black. Broad has conclusively shown that, on *this* definition of probability, use of the Laplacean formulae cannot be extended from the artificial to the 'natural' cases of inductions about substances and events. This, I think, is enough to refute, if not all, at least the great majority of the attempts to apply the Laplacean formulae to cases in nature.

The Laplacean formulae are also discussed by Broad in [6]. We shall therefore have occasion to return to the subject later.

B. *The hypothetical method*

Under this heading Broad considers, how the probability of an arbitrary proposition h is affected by the confirmation of some logical consequences, c_1, c_2, *etc.* of it. An important special case is, when h is a law (hypothesis, generalization) and c_1, c_2, *etc.* particular instances of the law.

Let the factual premiss of a certain inductive argument be that a certain hypothesis h has been confirmed n times in succession. How can this factual premiss, $c_1 \& \ldots \& c_n$, be used to probabilify the conclusion h?

About this Broad proves[16] something which may be called the Fundamental Theorem of Confirmation. It states, loosely speaking, that the probability of h increases each time the number n of successive confirmations increases, provided that (a) the probability of h prior to any confirmation is >0, and (b) the probability of the new confirmation relative to the previous confirmations is <1.

Substantially the same theorem was proved by Keynes in his *Treatise on Probability*.[17] But whereas Broad's proof is brief and elegant, Keynes' is unnecessarily complicated for the purpose. From the formula deduced by Broad, but not from the formula given by Keynes, it can be immediately seen that the increasing probability of the hypothesis approaches 1 as a limit, if and only if the probability of n successive confirmations approaches the initial probability of the hypothesis. Keynes states the condition in an equivalent but much less perspicuous form. This takes him into a very obscure discussion of the question whether the condition could be fulfilled in nature. The problem of convergence towards 1 is not discussed by Broad.

The more the probability of a new confirmation relative to the previous confirmations falls short of 1, the more does this new confirmation contribute to increase in the probability of h. 'This', Broad observes[18], 'is the precise amount of truth that there is in the common view that an hypothesis is greatly strengthened by leading to some surprising consequence which is found to be true.'

The Fundamental Theorem of Confirmation, one might say, makes the probabilification of inductive conclusions through factual premisses depend upon two additional premisses about extreme probabilities. The first additional premiss concerns the initial probability of the conclusion. The second is about a certain probability-relation among the factual premisses.

Of both the additional premisses it holds good that no assurance of their truth can be obtained from necessary principles of probability theory or pure logic. Their truth-ground, if any, must be some feature of the actual physical world. What it might be in the case of the second premiss, Broad does not consider. But in the second part of [1] he deals very elaborately with the possible truth-grounds of the assumption that laws for the explanation of observed phenomena possess, prior to confirmation, a finite probability.

Of the relation, finally, between Induction by Simple Enumeration and the Hypothetical Method Broad says[19] that the former is a special case of the latter. Now this may be true under some definition of the two methods. But it does not hold for Broad's conception of them in [1]. It is an essential feature of what Broad here calls the Hypothetical Method that the conclusion h should *entail* the factual premiss $c_1 \& \ldots \& c_n$. Only on this condition can the Fundamental Theorem be proved. But in that which Broad here calls Induction by Simple Enumeration there is no corresponding relation of entailment between conclusion and factual premiss. This makes a difference between the methods which, I think, is of some relevance to the discussion of their material presuppositions which follows in the second part of [1].

3. In several places[20] Broad makes apologetic remarks about the unsatisfactory nature of the discussion in the second part of [1] and about the doubtful character of the suggested ideas. It must, I believe, be admitted that there is much in this paper which is obscure, but also that it contains a wealth of interesting material. Much of the material is more relevant to the metaphysics of nature than to the logic of induction. In order to see clearly what is relevant in it to inductive logic, I think it is useful to distinguish the following three questions:

i. Which are the assumptions about nature which Broad thinks are necessary in order to make inductive conclusions, formulated in terms of probability, follow logically from factual premisses and formal principles of probability theory?

ii. Do these assumptions really serve their purpose of making inductive arguments formally valid?

iii. Are these assumptions true?

Broad does not keep these questions sharply apart, which makes the discussion of them somewhat difficult. But I think we may eventually assess the value of his arguments to all of them. I shall discuss the three questions in order.

i. We found above that use both of Induction by Simple Enumeration and of the Hypothetical Method requires certain 'material' assumptions about probabilities, and we tried to state which these assumptions were. In the case of enumerative induction an assumption of equi-probability of alternatives is required. In the case of hypothetical induction, we need two assumptions about not-extreme probabilities. Actually, the discussion in the second part of [1] touches only upon *one* of these three assumptions, *viz.* the assumption that laws have a finite initial probability. Thus the discussion, even if it were in itself quite satisfactory, would necessarily fall short of achieving *everything* that is needed to make inductive conclusions follow logically with probability from their premises.

Of the assumption about nature which is to assure a finite probability to laws, Broad says [21], that it must be an assumption which favours laws, *i.e.* propositions of the type 'All *S*'s are *P*' or 'No *S*'s are *P*' at the expense of statistical generalizations, *i.e.* propositions of the type '*p*% of the *S*'s are *P*'. This, I think, is false. For, from the point of view of the Fundamental Theorem of Confirmation, laws and statistical generalizations do not count as alternative hypotheses. This follows from what was just said about the relation between enumerative and hypothetical induction (in Broad's sense). Thus in discussing the problem of finite initial probabilities, we need not look round for a principle which *favours* laws as against statistical correlations. But we certainly must look for some principle *about* laws.

Here the idea of Limited Independent Variety suggests itself to the inquirer. If we had some assurance that, loosely speaking, the number of alternative laws for the explanation of a given phenomenon were necessarily finite, then there *might* also be assurance that the initial probability of any given one of these alternatives is not vanishingly small. This possibility Broad may be said to explore in [6]. In [1] he takes another route. It starts from considerations about the meaning of the idea to which philosophers have referred by some such name as Uniformity of Nature.

How is the 'axiom' about Uniformity of Nature – Broad calls it *Unax* – to be formulated? Broad first suggests [22] the following formulation:

if any individual a has the property ϕ and the property ψ (e.g., is a swan and is white) then there is some property χ other than whiteness (*e.g.*, that of being European) which is possessed by a, and such that everything that is both ϕ and χ (*e.g.*, is a European swan) is also ψ (*e.g.*, is white).

This seems to come very near the core of certain familiar views of universal determination in nature. Yet, as Broad points out [23], this formulation cannot be final. It is deficient in at least two respects:

Firstly, it is too general. It puts no restrictions upon the nature of the properties ϕ, ψ, and χ. It is plausible enough to think that the whiteness of a swan is due to the presence of some property (or conjunction of properties) such that any swan with this property is white. But how many people would hold that the 'swanness' of a white object is due to the presence of some property (other than the defining criteria of being a swan) such that any white object with this property is a swan?

Secondly, *Unax* in the above formulation entitles us to conclude from the conjunctive occurrence of two properties ϕ and ψ in a single individual a to the existence of some general law connecting these two properties. And this seems too bold a conclusion. In most cases, we would wish to witness quite a number of conjunctive occurrences of ϕ and ψ before we would feel justified in suspecting the existence even of *some* law connecting them. 'Yet', Broad says [24], 'it is very difficult to see what principle about *nature* there could be which makes *number of observed conjunctions* relevant at just this point.'

Broad thinks [25] that both these difficulties can be overcome, if *Unax* is modified in the following way: We demand that ϕ should be a property defining a Natural Kind.[26] The modified principle then says that, if the individual of the kind defined by ϕ possesses a further property ψ, then there is some property χ such that all members of the kind, which have χ, also have ψ. (χ may be the same as ϕ.) The uniformity principle, thus modified, Broad calls *Unaxk*.

ii. We now move on to the question, whether *Unaxk* can help to make inductive conclusions, formulated in terms of probability, follow logically from factual premisses and necessary principles of probability theory. As already explained (p. 237), the only help which *Unaxk* might give, is to assure a finite initial probability to laws of nature.

The way in which, according to Broad[27], the theory of natural kinds may be relevant to considerations of inductive probability can be briefly described as follows:

A natural kind is a region in nature containing a 'blob'.[28] The question then is whether 'this habit of heaping instances round a comparatively few possible states is typical of nature as a whole'.[29] Now suppose, for the sake of argument, 'that nature as a whole really distributes its instances uniformly among possible sorts'.[30] Then we shall have to go on to assume 'that the position of the human race is in some way wildly abnormal so that the parts of nature which have fallen under its observation have been utterly non-typical of the whole.'[31]

The last assumption would mean, either that human beings happen to live in a spatio-temporal region which is very unlike the rest of the universe, or that some limitations in our perceptive powers or interests have prevented us from noting all but instances of a few possible sorts.[32] After examination[33], the second alternative is rejected. We are left with the first alternative.

Now if the heaping of individuals about kinds be a peculiarity of a small section of the universe, whilst elsewhere the distribution is nearly uniform, it is highly unlikely that human observers will have happened to fall just into this part of the universe. The larger we suppose the universe to be compared with the part of it which has this peculiarity the less likely it is antecedently that – human experience should have fallen totally inside this peculiar region.[34]

The objection may be advanced[35] that the human race arose from definite causes in a definite part of the universe and that therefore the talk of its being 'shot at random' into the world is nonsense. Broad tries to show[36] that this objection is invalid. The talk of causes of the origin of the human race presupposes that the part of nature which has fallen under human experience is not peculiar in its nomic structure, but that causation operates outside this part in the same way as it operates inside it. The objection thus begs the question. I am not quite convinced by Broad's argument, but shall not subject it here to closer examination. Broad's conclusion anyway is that it is highly improbable that the general characteristic of confinement to kinds, which we have noticed, extends but slightly beyond the limits of human experience. We thus seem justified in disregarding the possibility that this characteristic of the experienced world does not extend beyond it, as an argument against induction.[37]

How, if at all, is Broad's probability-argument about induction and natural kinds relevant to the question of initial probabilities of laws of nature and thus to the question of a correct formulation of inductive arguments in terms of probability? There is no answer to this question in Broad's paper, and the question itself is not quite clearly presented.[38] In one place he says[39], quite rightly I think, that this argument may be used to probabilify the *general* view that there are kinds and nomic connexions in nature, rather than used to probabilify any *special* generalizations about such and such kinds or such and such causal relationships. Thus he comes himself near to admitting that his argument cannot, in combination with the Fundamental Theorem of Confirmation, be used to raise the probability of any single inductive conclusion appreciably above the zero-level.[40] I think that this admission is necessary and that *Unax* and the theory of natural kinds cannot – unless strengthened very much beyond the content Broad gives to these ideas – be used for securing finite initial probabilities to (some) laws of nature and thus for satisfying *one* of the conditions for increase in the probability of inductions through confirming instances.

iii. The answer to the second of our three questions above is thus, I believe, negative. The general principles about the constitution of the physical world which Broad discusses in [1] do not serve the purpose of making inductive conclusions, formulated in terms of probability, follow logically from factual premises and necessary principles of probability theory.

It follows from this that the question of the *truth* of the general principles is not directly relevant to the discussion in [1] of the nature and justification of inductive inferences. By this I mean that, even if we could prove these principles to be true, we should not, from what we are being told in [1], know how to use them to probabilify inductions. This I shall take as an excuse for not discussing here the problem of truth in question. But I do not wish to say that the theory of natural kinds in [1] could not be further elaborated so as to become relevant to the problem of a 'rational reconstruction' of inductive arguments in probabilistic terms. And, quite apart from this, the theory itself is of great interest.

I shall conclude my examination of [1] by briefly drawing attention to two points in Broad's theory of kinds, substances, and causation, which, though scarcely directly relevant to the problem of probabilifying inductive conclusions, yet seem to me very important. I regret that I cannot possibly do justice to the wealth of ideas contained in Broad's own discussion of the topics.

(a) The assumption of the existence of kinds of substance involves an assumption of the existence of substances. In speaking of the species *swan* we assume that there are individual substances, *i.e.* persistent physical things, which may be identified as swans. But what is our evidence that there are such things? A persistent thing may, somehow, be regarded as a complex of its 'states'.[41] Such states are what we observe of the thing. Now saying of two observed states, separated in time, that they are states *of the same thing* involves rather complicated assumptions about the existence of unobserved intermediate states, related to one another and to the two observed states in a certain way. Our evidence for these assumptions consists of observations on series of states of other substances *of the same kind*. Thus the three notions of natural kinds, of individual substances, and of plausible inductions are seen to be interdependent. Kinds presuppose substances; the evidence for substances is inductive; the warrant of the plausibility of inductions are kinds.

I am not sure that I quite understand Broad's argument about the relation of the assumption of kinds to the assumption of substances and particularly not about the relation of substances to their states. But the circularity as regards the basis of induction, which Broad points out, seems to me to be unavoidable. It does not follow that the circularity is necessarily of a vicious kind, nor does Broad say that it is. As far as I can see, he does not investigate the possible consequences of the circularity in question to the problem of justifying induction.

(b) 'The world as it presents itself to superficial observation,' Broad says [42]

fulfils to a highly surprising extent the condition of consisting of permanent substances of a few marked kinds. – But it does not fulfil it altogether. The position is that it fulfils it so well as to raise the expectation that a modification of the definition of permanence and of kinds, which shall be in the spirit of the original definitions, can be found, and that with this definition the universe will *strictly* consist of permanent substances belonging to a few ideal kinds.[43]

The required modification is accomplished, when we make use of causal notions to account for (α) why the properties of the states, a series of which constitutes a thing, do not always remain constant throughout the series but sometimes come into being, vary, and pass away, and (β) why contemporary states of different series, which constitute substances of the kind, are not exactly alike but deviate more or less from what may be called the 'norm' or 'ideal' of the kind.[44]

Thus by an extension in terms of causality, of the notions of substance and kind we may restore the ideas of permanence of substance and ideality of kinds. 'The permanence of first-order properties and their exact similarity among all instances, which first suggested kinds and permanent things, breaks down; but it is replaced by *permanence of laws*'.[45]

There is also another way, in which causation becomes involved in the notion of a kind, *viz.* when a kind is defined (partly) in terms of causal properties of substances. This happens, for example, when it is regarded as part of the definition of silver that it is the kind of substance which gives a white insoluble compound with chlorine.[46]

Thus the notions of substance, kind, and causation can truly be said to form 'parts of a highly complex and closely interwoven whole and any one of them breaks down hopelessly without the rest.'[47]

That there is a connexion between the three notions mentioned can hardly be denied. The connexion, moreover, is probably of such a character that inductions about properties of kinds of substance (such as 'all crows are black') depend for their formal validity (either in Broad's sense in [1] or in some related sense), not only on some general principle about natural kinds (such as U_{naxk}), but also on some or several general principles about causation between events. What precisely this dependence is, cannot, I think, be seen from Broad's paper. But what he has to say about the matter is sufficiently interesting to be a challenge to others to try to make these intricate connexions in the conceptual groundplan of nature more perspicuous.

II

1. In [6] Broad takes the same view as in [1] as regards the correct formulation of problematic inductions, *viz.* that their conclusions should be stated in terms of probability.[48] This opinion was criticized above (pp. 231–232).

If from a premiss of the form 'All observed S's have been P' we draw a conclusion (as to the probability of a proposition) of the form 'All S's whatever are P', we have what Broad calls[49] a Nomic Generalization. If from a premiss of the same form we draw a conclusion of the form 'The next S to be observed will be P' we have a Nomic Eduction. If from a premiss of the form 'A certain proportion of the observed S's have been P' we draw a conclusion of the form 'A certain proportion of the total number of S's are P' we have a Statistical Generalization. And if from a premiss of the same form we draw a conclusion of the form 'The next S to be observed will be P' we have a Statistical Eduction.

The term 'eduction' Broad has from W.E. Johnson. The distinction between eduction and generalization is certainly useful. But I am not quite happy with Broad's (and Johnson's) way of making it. Since the matter is of more than terminological interest, I shall dwell upon it for a moment.

Suppose the conclusion of an inductive argument were of the form 'The n next S's will be P'. How shall this case be classified? It would seem that, if n equals 1, we have a case of eduction, and if n equals the total number of unobserved S's we have a generalization. Further, if n is greater than 1 but less than the total number of unobserved S's, we have a third case which is *neither* eduction *nor* generalization, and if n equals 1 and the total number of unobserved S's is also 1, we have a fourth case which is *both* eduction *and* generalization.

Broad does not make a sharp distinction between an inductive conclusion which applies to an (at least 'potentially') infinite multitude of unobserved cases and one which applies to a numerically restricted multitude. Instead of 'infinite multitude' we may (here) also say 'open class' and instead of 'numerically restricted multitude' we may also say 'closed class'. This distinction is far from unproblematic. There are some indications that Broad has wished to avoid the problems relating to the notions of generality and infinity with which this distinction is intimately connected. Yet I think that this distinction is necessary for the purpose of classifying inductive conclusions. And I would myself reserve the name 'generalization' for inductive conclusions about open classes, and use the term 'eduction' for inductions about closed classes.

Now, as far as I can see, the theory of inductive probability which Broad develops in [6] and which is a variant of the doctrine of inverse probability applies only to inductive conclusions about *closed* classes. This theory is thus a theory of what I would call the probability of eductions, as opposed to (genuine) generalizations.

The theory of generators, on the other hand, which Broad also develops in [6] and which may be characterized as a further elaboration of some suggestions made by Keynes, is a theory about the logic of nomic connexions in nature. And it is at least highly plausible to think that generalizations about such connexions are propositions about *open* classes. This makes it difficult to see how the theory of generators could be relevant to Broad's theory in [6] of the probability of inductions, although the purpose of the former may be said to be to warrant certain assumptions about nature which turn out to be necessary in the latter. As we shall see later, Broad does not succeed in linking the two theories with one another. I would myself doubt whether such a link can be found. I should venture to maintain that the theory of generators is relevant neither to Broad's nor to any other version of the doctrine of inverse probability (Bayes's Theorem, Laplace's Rule of Succession, *etc.*) but to quite another type of theory of inductive probability. This other type of theory deals with the probability of generalizations relative to confirming evidence (instances) entailed by the generalization. This theory, though founded by Broad in [1], is not treated by him in [6]. Thus [6] consists in fact of two parts which are not relevant to one another.

2. As in [1], Broad first considers the drawing of counters from a bag *without* replacement. He produces a very handsome derivation, which makes no use of integration, of the formula $\frac{m+1}{m+2}$ for Nomic Eduction and $\frac{m+1}{n+1}$ for Nomic Generalization. These are called Laplace's First and Second Rules of Succession. The proof requires the two assump-

tions of equi-probability mentioned in [1] and discussed above on p. 233.

The assumption that all proportions of, say, red counters in a bag are initially equally probable, which Broad accepted in [1], is now rejected on the ground that several different so-called *constitutions* of the bag may answer to one and the same *proportion* of red counters in it.[50] Broad speaks of it as the false[51] assumption of equi-probability. As an alternative to it he considers the following assumption which he thinks is true[52]: Any individual counter is as likely to have any specific colour as any other. (The number of distinguishable colours is assumed to be v.) Under this assumption the Laplacean formulae cannot be derived. Instead we get for Nomic Eduction the, somewhat disappointing, result that the probability that the *next* counter will have a certain colour is not at all affected by the fact that all drawn counters have had this colour.[53] For Nomic Generalization, however, we get a formula showing that the probability that *all* counters will have a certain colour increases with the number of already drawn counters which have been found to have this colour.[54]

In calling the one assumption of equi-probability false and the other true Broad evidently thinks that the first can be shown to be an invalid and the second a valid application of a Principle of Indifference. In [1], moreover, he regarded this principle itself as an *a priori* truth of probability-theory.[55] In [6], he does not mention it among the formal principles of probability and logic, necessary for his proofs.[56] I regret that Broad has nowhere discussed in detail this most important and controversial idea in the philosophy of probability.[57]

I do not believe myself that there is any way of *proving* or *disproving* either assumption of equi-probability. The distinction between constitutions and proportions gives no conclusive evidence against the Laplacean assumption. The distinction only shows that there is no *obvious* way of 'splitting up' the situation into unit-alternatives of equal 'weight' even in the seemingly simple case of proportions of counters in a bag.

It is noteworthy that Broad does not consider, as an alternative to the rejected assumption of equi-probability, the assumption that all constitutions of the bag are initially equally probable. If I am not mistaken, the assumption of equi-probability of constitutions would lead to the same, disappointing, result in the case of Nomic Eduction as does

Broad's assumption about equi-probability of colouring, *viz.* that no amount of favourable evidence in the past can affect the probability that the next ball is, say, red. This 'eductive inefficiency' of the assumption of equi-probable constitutions, incidentally, has been used by Carnap[58] as an argument *for* (a modified form of) the original Laplacean assumption of equi-probable proportions.

Broad goes on to consider some probability-problems in connexion with the throwing of a counter with two faces. These problems are equivalent to problems of drawing counters from a bag *with* replacement. I shall here skip the mathematical part of the discussion and proceed to the important notion of *loading*. We say that the counter is loaded to a degree s in favour of red, if, and only if, the antecedent probability of its turning up red would be s for anyone who knew in detail how it was constructed.[59]

3. The notion of loading is not peculiar to problems on throwing counters. It is present also in problems on drawing with or without replacement. In stating the above First Premise of Equi-probability we assume that the counters have an equal load in favour of being drawn.[60]

Now Broad makes an extremely interesting suggestion about the evaluation of a load or, as I should prefer to call it, antecedent probability. If I understand him rightly, he wants to say that *any* such evaluation has, in the last resort, to rely on some assumption of *equi-probability*. We could not, for example, evaluate the probability of a counter falling with the red side upwards unless we knew the antecedent probabilities of its striking the table at each of the possible angles. And these antecedent probabilities, Broad goes on to say[61], could not be evaluated without some assumption about equi-probability. This assumption could be, *e.g.*, that it is equally probable that the counter will strike the table at any given angle as at any other.

'The notion of loading,' Broad says[62], speaking of throwing a counter, 'is the notion of a constant cause-factor which operates throughout the whole series of throws and co-operates with other and variable cause-factors to determine the result of each throw.' I am not sure in what sense the term 'cause-factor' should be understood here. The load, I should have thought, is a certain antecedent probability. Is this probability *itself* a cause-factor? Or does the cause consist in some features of the physical world (other than probabilities), which may be held "responsible" for the load? (Such features could be, *e.g.*, properties of the counter and conditions under which it is being thrown.)

The question is of some importance, since Broad now goes on to maintaining that 'every inductive argument, whether it be a nomic generalization, an eduction, or a statistical generalization, equally pre-supposes the notion of causal determination'.[63]

If, for the sake of argument, we assume that every inductive conclusion in terms of probability is dependent upon the notion of load, *i.e.* upon antecedent probabilities, *and* that every load has ultimately to be evaluated in terms of equi-probabilities, then the following question arises: Does Broad's statement about the dependence of induction upon causation amount to anything *more* than the said dependence of induction upon antecedent probability *and* of antecedent probability upon equi-probability? If the answer is in the negative, then I think that Broad's formulation above somewhat overstates the nature of the dependence in question. I hope Broad could illuminate this point in his reply.

I shall not here restate or examine the Fundamental Causal Premise upon which, according to Broad, any inductive argument (in terms of probability?) rests[64]. This premiss is, as Broad observes[65], not the same as the Law of Universal Causation.

4. After his discussion of the causal presuppositions of the theory of inductive probability, Broad returns to the Rule of Succession. The probability, which the Laplacean rules confer upon eductions and generalizations could presumably be increased, if we could replace the Laplacean assumption of equi-probability of degrees of loading by some other assumption which favours a high degree of loading (in favour of a certain result in the experiment). This other assumption Broad calls an Assumption of Loading. 'It is required', Broad says[66], 'not to validate inductive arguments as such, but to validate the claims of *some* of them to produce *high* probabilities.'

As far as I can see, it is one of the objectives of the Theory of Generators to justify such Assumptions of Loading.[67] Exactly *how* the theory is supposed to do it, I have not however been able to figure out. (*Vide infra.*)

5. When from the 'artificial' cases considered in sections 2–4 we pass to 'natural' cases such as inductions about the properties of substances or the causal concatenation of events, a number of differences should be noted which affect the conditions of applicability of the formulae of inverse probability. Broad's comments in [6] on this topic are similar to his comments on it in [1].[68] The following three points are emphasized:

i. The number of members of a natural class (such as the class of all swans) is *unknown*. Broad says[69] that 'it is almost certainly very great as compared with the number that have been observed up to any given moment'. As a consequence of this the value $\frac{m+1}{n+1}$ will be vanishingly small, and the Laplacean formulae will anyway be useless for assessing the probability of nomic and statistical generalizations. – I would myself have gone a step further and said that the number of members of a natural class is 'potentially infinite' and that we therefore have no right to assume even that it is *some* finite number n. The Laplacean formulae for nomic and statistical generalizations have no *application* to natural classes.

ii. The same object (*e.g.* a swan) may be observed several times without our noticing the identity. Thus we may be led to thinking that the number m is greater than it really is. This will lead us to over-value inductive probabilities.

iii. Our observations are from a limited region in space and time. This means that the assumption breaks down, according to which every individual in the class has an equal chance of being observed by us. On the function and necessity of this premiss of equi-probability we made some comments above on p. 233f.

The upshot of all this is, according to Broad[70], that if the application of inductive arguments to nature is to lead to reasonable probable conclusions, then

(a) 'we must have some reason to believe that something analogous to "loading" exists in nature, and that certain kinds of "loading" are antecedently more probable than others', and

(b) we must somehow get over the objection which follows from iii above as regards equi-probability.

I must confess that I do not see how Broad's demand under (a) is linked with his observations on the difference between the artificial and the natural cases. Nor can I see how the Theory of Generators and Limited Independent Variety, which Broad now proceeds to develop, satisfies either demand. We now pass on to an examination of this theory.

6. The germ of the Theory of Generators is found in the second half of the part on induction and analogy in Keynes' *Treatise*.[71] Its development by Broad, however, is an essentially original and important contribution of his to logic. It is worth a much more detailed presentation and discussion than can be given to it in this essay. The subject is rather technical. I hope someone would take it up for further development and I believe that this should happen within the framework of a Logic of Conditions. This framework was developed and used by Broad himself in [7] for his theory of so-called demonstrative induction which we shall examine in the next section of the present paper.

The Theory of Generators is based upon certain assumptions which may – with some simplification – be stated as follows[72]:

Let us assume that we are given two mutually exclusive sets of characteristics (properties). We call them a set of generating and a set of generated characteristics respectively. The first contains n and the second N (logically and causally independent) members. The number n is finite and smaller than N. Every single member of the second set is 'generated' by some member or conjunction of members from the first set. This means that whenever the member or conjunction of members from the set of generating properties is present, then the member in question from the set of generated properties will be present too. One and the same member from the set of generated properties may be generated by more than one member or conjunction of members from the set of generating properties. If this actually is the case, we say that the member in question of the set of generated properties has a Plurality of Generators.

We shall call the assumptions underlying the Theory of Generators by the name of the Principle of Limited Variety.[73] We may distinguish two forms of the Principle: a stronger form which excludes and a weaker form which admits Plurality of Generators.

Broad does not state explicitly, whether it is essential to the Principle

249

of Limited Variety that the number N too should be *finite*. But in at least one place his argument presupposes that N is finite.[74] The assumption of a finite N seems to me to lessen considerably the *prima facie* plausibility of the Principle of Limited Variety as a proposition of the Metaphysics of Nature.

The sense in which the Principle of Limited Variety may be said to *limit* variety in nature is worth some special comments:

Consider any true Nomic Generalization which holds between two mutually exclusive sets of generated characteristics.[75] An example could be the proposition that anything which has C_1 and C_2 and C_3 also has C_4 and C_5. C_1, C_2, and C_3 are called subject-factors, C_4 and C_5 predicate-factors. Now assume that C_1 and C_2 and C_3 are, in the above sense, generated by G. Then, by laws of logic, C_4 and C_5 are also generated by G. Thus the truth of the Nomic Generalization in question may be attributed to the generating capacity or, as Broad calls it[76], *fertility* of the generator G. Since, on our assumption, the number of generating characteristics and conjunctions of characteristics is *finite* (it is $2^n - 1$), it follows that *all* Nomic Generalizations which hold true for sets of generated characteristics may be thus attributed to the generating capacity of a limited, finite number of generators.

Unless we assume that N too is finite, there is nothing to exclude a generator from having *infinite fertility*, i.e. from having the capacity of generating an infinite number of generated characteristics. (Indeed, if N *is* infinite, there must be at least one generator of infinite fertility.) Thus the Principle of Limited Variety, without the assumption of a finite N, does not entail that there is a finite number of irreducible nomic connexions (between generating and generated characteristics) in nature.

We now come to the question, how the Theory of Generators is relevant to the antecedent probability of Nomic Generalizations (for generated characteristics). Broad conducts two different arguments to show that this antecedent probability will be greater than 0. The second argument leads to a somewhat stronger conclusion than the first.[77]

According to the first argument[78], the antecedent probability of an arbitrary generalization can be expressed as a product of two probabilities. The one is the probability, given the Principle of Limited Variety, that the (combination of) subject-factors of the generalization are generated by (the conjunction of) exactly r members of the set of generating characteristics and the (combination of) predicate-factors by (the conjunction of) exactly s members of the set mentioned. The other is the probability that the s generating factors required by the predicate are wholly contained among the r generating factors required by the subject. Broad thinks that the product-probability will be greater than 0.[79] This commits him to holding that both factors of the product are greater than 0. The second factor he identifies with the ratio of the number of ways of choosing s things out of r things to the number of ways of choosing s things out of n things. One gets the impression that he regards this identification as something unproblematic – which it hardly is. The ratio in question is, on our assumptions, certainly greater than 0. The first factor Broad does not attempt to evaluate numerically. He explicitly says [80] that he does not see how to do it, and mentions some difficulties which seem to me to be very much to the point. And, for all I can see, he has not even proved that the factor must be greater than 0. For *this* reason already Broad's first argument seems to me to be inconclusive.

The second argument requires some interesting Lemmas. Of the relative values of n and N Broad proves [81] that if $n \geq N$, then there *may be no* true generalization (for generated characteristics), and if $n < N$ then there *must be some* true generalization. Now it is part of the assumptions that $n < N$. If, moreover, some of the possible Nomic Generalizations (for generated characteristics) are true, then, by laws of logic, some *such* generalization must be true, the subject of which is the conjunction of all but one of the generated characteristics and the predicate of which is the one remaining generated characteristic. There are in all N Nomic Generalizations of this description. From this Broad concludes that the antecedent probability of any of them will be at least $1/N$. If N is finite this fraction is greater than 0.

It is to be noted that the second argument presupposes that N is *finite*. (The first argument is independent of this assumption.) It further presupposes that the antecedent probability of a generalization can be linked with a ratio of true generalizations. (In this respect the first and the second argument resemble each other.) The nature and justification of the link is not made clear.

The upshot of the matter seems to me to be that Broad has *not* been completely successful in showing how the Theory of Generators assures a finite antecedent probability to Nomic Generalizations. Assume, however, for the sake of argument, that he had succeeded in showing this. What further implications would follow for the theory of the probability of inductions?

As far as I can see *nothing* would follow which is relevant to the theory developed by Broad in the first half of [6], *i.e.* to the probability-theory of what he in [1] calls Induction by Simple Enumeration. But *something* important would follow for the probability-theory of what Broad in [1] calls the Hypothetical Method. It would follow that *one* of the two conditions which are necessary, if the probability of a Nomic Generalization is to be increased through confirmation, is fulfilled. But it would not follow that the second condition is fulfilled.

The reasoning has so far been conducted on the assumption of the strong form of the Principle of Limited Variety which excludes Plurality of Generators. Later Broad indicates [82] how the reasoning may be modified so as to apply also under the weaker assumption admitting Plurality of Generators. Then the question of evaluating probabilities on the basis of combinatorial calculations becomes still more tricky and dubious.

We shall not here discuss at length the plausibility of the assumption that properties which occur in nature can be divided into exclusive sets of generating and generated characteristics. Broad shows [83], very elegantly I think, how we may *dispense* with this division altogether and yet retain everything that is essential to the Principle of Limited Variety. Instead of postulating the existence of a distinct set of generating properties, we assume that the one set of (generated and generating) properties falls into a finite number of what Broad calls [84] *coherent sets*. A set of characteristics is said to be coherent, if no member of the set can occur without all the remaining members occurring. This simplifies the Theory of Generators. But it does not help us to clear up the problem concerning its relevance to the probability of inductions.

It is clear that – whether or not we wish to dispense with the assumption of a distinct set of generators – we could not profitably discuss the plausibility of the Principle of Limited Variety unless we first made clear, to what sort of 'characteristics' or 'properties' this principle is intended to apply. It may well be the case that for some kinds of characteristic the principle has a certain *prima facie* plausibility, whereas for other kinds it has none.

Broad thinks[85] that the characteristics ought to be *determinables*. It would be absurd, he says[86], to suppose that the members of the set of generating properties are determinates. The reason he gives is that any determinable may have an infinity of determinates falling under it. And if the generating properties are determinables, the properties which they (or their conjunctions) generate must be determinables too. This, in Broad's view, follows from the nature of determinables and determinates.

The restriction of the characteristics, to which the Theory of Generators may apply, to determinables diminishes its importance to inductive logic. The theory could then, if at all, only serve to probabilify generalizations of the type which Broad in [7] calls Laws of Conjunction of Determinables. Of them Broad says[87] that 'only the most backward sciences are content with such generalizations'.

Neither in [6] nor in [7] does Broad define the notions of determinable and determinate. I think he intends to use them in the same sense as W. E. Johnson. But the precise meaning of the terms with Johnson is not clear to me. (Indeed, I find Johnson's use of them ambiguous and confused.) I have therefore not been able to form an opinion of the seriousness to inductive logic of the alleged restriction of generator-properties to determinables. I hope Broad could illuminate the point in his reply. Of the notions of determinable and determinate we shall have to make a few more remarks in connexion with the discussion of [7].

7. The Principle of Limited Variety was thought by Broad to be an ontological precondition, 'if inductive arguments are ever to be able to establish reasonably high probabilities'. Leaving aside the problem of necessity and sufficiency of this condition, we raise the question: Can we know, whether it is in fact fulfilled? Broad calls this the *epistemic* question.[88]

Broad regards it as excluded that we could know the Principle of Limited Variety with *certainty*. But he thinks that an argument can be conducted in favour of the principle's *probability*.

I shall not reproduce here the formal part of the argument. Its material content consists of two probability-assumptions.

253

The first rests upon an analogy:

> We do know that we can actually construct out of simple parts of the same nature complicated structures which behave in very different ways, *e.g.*, watches, motor-cars, gramophones, *etc.* The differences in observable behaviour are here known to be due simply to differences in arrangement of materials having the same properties; and these materials, and the structures formed of them, are parts of the material world. Relative to this fact it does seem to me that there is a finite probability that the variety of *material* nature at any rate, should arise in the same way.[89]

Broad observes[90], that there is no similar ground for believing in a Principle of Limited Variety as far as *mental* phenomena are concerned.

The material basis of the second probability-assumption involved in Broad's argument is the observed fact 'that there is a great deal of recurrence and repetition in nature'.[91] 'Now, if the Principle of Limited Variety were true, there would be recurrence and repetition in nature; whilst if it were not, there is very little reason to expect that there would be.'[92]

I do not find Broad's probability-argument either from the analogy with machines or from the observation of regularity in nature very convincing. A conclusive test of the argument's validity would, however, require that the argument itself is first given a much more rigorous formulation than is the case in Broad's paper.

Someone may think that the question of the truth or probability of the Principle of Limited Variety is anyway greatly diminished in importance by the fact that the principle can do, if anything at all, even less than Broad thought in [6] to strengthen inductive arguments. Its role as an ontological precondition of induction is doubtful, to say the least.

But quite apart from its relevance, if any, to induction, the idea of limited variety in nature seems to me highly interesting. If I may conclude by expressing a personal opinion on the matter without giving sufficient reasons for it, I should say that the interest does not lie in the question of the *truth* of the idea so much as in the problem of its *formulation*. As we have seen, it is not obvious which assumptions should be regarded as essential to the Principle of Limited Variety (pp. 249–250 above) nor is it obvious which aspect of reality constitutes its field of application (pp. 252–253 above). To consider various alternatives to the solution of these problems is to deepen our insight into the conceptual network of thinking about nature and is *therefore* already a major task of a Natural Philosophy or Metaphysics of Nature. It is my belief that the nearer we come to accomplishing this task, the less will the problem of truth worry our minds. The craving turns out to be for clarity – not for truth or probability. [...]

III

1. In the widest sense, a *demonstrative induction* may be defined as follows: '*P* and i_1 and ...i_n, therefore *G*'. *G* is a generalization, of which i_1, etc. are instances. *P* is the so-called supplementary premiss. The argument is conclusive, *i.e.* the generalization follows logically from the conjunction of some of its instances and the supplementary premiss.

Following W.E. Johnson, Broad[93] gives a more restricted definition. He requires that *P* be a hypothetical proposition. The consequent of this hypothetical proposition is *G*. The antecedent is either an instance *i* of *G* or an existential (particular) proposition, entailed by *G*. The minor premiss is *i*.

An example of a demonstrative induction in Broad's sense would be: 'If hydrogen can be liquefied, then every gas can be liquefied. Hydrogen can be liquefied. Therefore every gas can be liquefied.'

The idea of a theory of demonstrative induction can be said to be implicit in Bacon's treatment of induction. It was, as far as I know, first formulated explicitly by Archbishop Whately[94], whom Mill followed in his often quoted *dictum* that 'every induction may be thrown into the form of a syllogism by supplying a major premiss'.

It is noteworthy, however, that when Broad later in [7] gives a formal re-statement of Mill's methods he does not state them in a way which answers to his own pattern of demonstrative induction. I do not myself see any plausible way of stating them thus. But I believe that they can be stated so as to answer to the more general pattern mentioned above. How this is to be done, I shall not discuss here. Therefore I think that the Johnson–Broad definition of demonstrative induction stands in need of modification, if it is to fit, as it is obviously intended to do, some of the most important types of conclusive reasoning in connexion with induction.

2. In the traditional theory of the so-called methods of induction the

notions of cause and effect hold a prominent place. As Broad observes[95], the word 'cause' is used very ambiguously both in ordinary life and in science. Sometimes the cause is being thought of as a necessary, sometimes as a sufficient, and sometimes as a necessary-and-sufficient condition of the effect.[96] The various kinds of condition are characteristically different in their logical properties. It may be shown that these properties are the only aspect of the notions of cause and effect which is relevant to the logical mechanism of *induction by elimination*. Elimination-theory, *i.e.* the main bulk of inductive logic in the tradition of Bacon and Mill, may therefore most conveniently be approached by way of the Logic of Conditions. To have inaugurated this approach seems to me to be Broad's greatest over-all contribution to inductive logic. His ideas place an old subject on a new basis and open interesting prospects for further investigation.

3. In [7], before proceeding to develop a logic of conditions, Broad observes that there are two different types of causal laws. The first he calls Laws of Conjunction of Determinables. For example: cloven-footed animals chew the cud. The second he calls Laws of Correlated Variation of Determinates. An example would be the law for gases, stating that $P = RT/V$. A science in an early stage mainly has to be content with laws of the first type. The more 'advanced' a science is, the more prominent are within it laws of the second type.

I believe that Broad is here aiming at a distinction which is very fundamental to the logical study of induction. But I am not quite satisfied that the distinction should be formulated as Broad does it here. My difficulty has to do, among other things, with some unclarities in the notions of determinable and determinate. (See above p. 253.)

These notions are obviously *sometimes* relative notions. 'Bird' is a determinable relative to 'raven' and a determinate relative to 'animal'. But it would be rash to maintain that there are no absolute determinates. And if there are, there are presumably also laws for their nomic connexions.

It seems to me, therefore, that we must accept as a third type of causal law Laws of Conjunction of Determinates. The fundamental dichotomy in the division of laws should – whatever be its precise nature – rather be formulated in terms of the contrast between *conjunctions* and *cor-*

related variations of characteristics than in terms of the opposites *determinable* and *determinate*. I shall in the following speak of conjunction-laws and variation-laws.

There are logical relations between the two types of law. Any variation-law, as far as I can see, entails a conjunction-law. (This seems at least to hold for Broad's conception of the two types.) For example: the formula $P=RT/V$ of correlated variations entails that pressure depends uniquely on temperature and volume. The entailed proposition may be called a conjunction-law. I understand that the first premiss in each of the four Johnsonian 'figures of induction', which Broad explains in the second part of [7], is a conjunction-law in this sense.

I agree that this type of premiss is necessary for the correct formulation of the figures. And it is probably right to think – as seems to be implied by Broad's description of the situation in the opening paragraph of the second part of [7] – that methodical search of variation-laws *presupposes* methods for the establishment of conjunction-laws. These methods evidently must be methods of elimination. (The reader is asked to consider, *how* we may convince ourselves that, say, the pressure of a gas depends on its temperature and volume and on no other factors.) To give a precise account of them within the framework of a Logic of Conditions is therefore a primary task of any satisfactory logical theory of induction.

4. Broad in [7] is, as far as I know, the first to have given systematic attention to the logic of the various notions of condition. Besides in [7] he has also treated the subject in [8] and, with considerable additions to his original theory, in [10]. It is not necessary here to enter deeply into the technicalities of the subject. I shall confine myself to a few observations of a general nature:

i. Broad defines[97] '*C* is a sufficient condition of *E*' as meaning 'Everything that has *C* has *E*', and '*C* is a necessary condition of *E*' as meaning 'Everything that has *E* has *C*'.

It is worth observing that the two notions are interdefinable. If *C* is a s.c. of *E*, then $\sim C$ (the absence of *C*) is a n.c. of $\sim E$ (the absence of *E*). That oxygen is necessary for life is equivalent to saying that absence of oxygen is sufficient to extinguish life.

This relation between the two types of condition, which follows from

Broad's definitions of them, appears to be quite in order. Broad, though recognizing [98] 'negative factors', does not mention it.

Another relation, however, which can also be deduced from the definitions, appears debatable. It is this: If C is a s.c. of E, then E is a n.c. of C. If rainfall is sufficient to wet the ground, then the ground becoming wet is necessary for rainfall. This sounds odd. The oddity, obviously, comes from the fact that the above definitions contain no reference to what may be called a *direction of determination*. In the realm of natural events and continuants, the conditioning factor is somehow *prior to* the conditioned factor. It is at least highly plausible to assume that this 'priority' has to do with the notion of temporal succession. (This certainly holds for what Broad calls [99] the popular-scientific notions of cause and effect.) In Broad's (and my) Logic of Conditions, however, the asymmetry between conditioned and conditioning factors has vanished.

Broad maintains [100] that, from the point of view of 'the logical manipulation' of the causal notions it is not necessary to pay attention to the idea of temporal succession. He *may* be right as far as the canons of induction by elimination are concerned. (I have certainly been inclined to think so myself in the past.) But I am sure that a fully satisfactory Logic of Conditions cannot be based on definitions of the various notions of condition in terms of universal material implication only. A logical theory which accounts for the notion of a 'direction of determination' is needed. And I believe that such a theory, once it has been created, can serve to give a much fuller account also to the logic of inductive methods than can the Logic of Conditions in its present form.

It should be added in fairness to Broad's theory that it avoids some of the oddities, which result from the definitions, by means of a device to separate determining factors or C-factors as Broad calls them from determined factors or E-factors.[101] This device, however, is purely notational. It does not help us to overcome any real difficulties in the theory.

ii. *Postulates.* The Logic of Conditions, as developed by Broad in [7], uses two 'postulates'. The first Broad calls the 'Postulate of Conjunctive Independence'. It states, roughly speaking, that the (simple) C-factors should be capable of independent presence or absence. Independence is here understood in the strong sense of both logical *and* causal independence. The second postulate is called the 'Postulate of

Smallest Sufficient Conditions'. It says that any E-factor has, in every instance of its occurrence, a Smallest Sufficient Condition. This means that 'whenever the characteristic E occurs, there is some set of characteristics (not necessarily the same in each case) such that the presence of this set in any substance carries with it the presence of E, whilst the presence of any selection from this set is consistent with the absence of E'.[102]

The second postulate is a greatly improved form of what most authors on induction and scientific method vaguely refer to under the name of Law of Universal Causation. It is clearly 'extra-logical'. It seems to me to be in the interest of the logical purity of the theory to develop the Logic of Conditions independently of this postulate. With its aid, but not without it, one can prove, for example, that any factor which is common to all the s.c.'s of a given E is a n.c. of E.[103] It is of some importance to notice that, contrary to what there may be a tendency to believe, the truth of this theorem does *not* follow from the definitions of the notions of condition and principles of pure logic alone. It would therefore give a clearer picture of the case, if the existential assumption on which this theorem depends were stated as part of the theorem itself and not as a postulate of the entire theory.

iii. *Complexity of Conditions*. The most serious insufficiency of Broad's Logic of Conditions in [7] is that it neglects disjunctive necessary conditions. In [10] Broad corrects his theory on this point. In [10] other important additions are also found.[103a] Among them is the introduction of the notions of *contributory condition* and *substitutable requirement*. If C_1 and C_2 are disjunctively necessary for E, they are called substitutable requirements of E. If they are conjunctively sufficient, Broad calls them contributory conditions. (In [8] he had coined the not very happy name 'relatively necessary conditions'.) A contributory condition which is a common ingredient in all the sufficient conditions of a phenomenon he calls an *indispensable contributory condition*. It is important not to confuse indispensable contributory conditions with necessary conditions.[104]

iv. *Plurality of Conditions*. Cause and effect are said[105] to stand in the relation of total cause to total effect when, roughly speaking, the effect is a conjunction of all the factors, of which the cause is a s.c. It follows that one and the same total effect may have several total causes, but that one and the same cause can have only one total effect. If, however, we allow disjunctive effect-factors (as I think we should do), then there is a sense in which we may speak of a plurality of effects too. For then we may have that C is *sometimes* followed by E_1 and *sometimes* by E_2. But even in this case the effect would be unique in the sense that C would *always* be followed by at least one of the two factors, E_1 and E_2. (Disjunctive effects are worth a closer scrutiny.)

Broad's statement[106] that, even under the Postulate of Smallest Sufficient Conditions, an effect may have *no* n.c. is of course false, if we admit conditions of a disjunctive form. It is of some interest to note that the postulate in question actually is equivalent to saying that every E has a *necessary-and-sufficient* condition, *viz.* the disjunction of *all* its s.c.'s.

5. Broad's formal statement of Mill's methods does not seem to me quite happy. Broad's reasoning, unlike Mill's, is entirely correct. But Broad, like Mill, fails to notice that there is an essential asymmetry between the two methods of Agreement and Difference, when used for the search of conditions of given phenomena. The Method of Agreement is (primarily) fitted for the task of finding the necessary conditions of a given phenomenon. In Mill's terminology; finding the effect of a given cause. Or as Broad put it[107]: finding *of what* a given phenomenon is a sufficient condition. The Method of Difference again is fitted for the converse task of finding the sufficient conditions of a given factor. In Mill's terminology: the cause of a given effect. In Broad's: finding *of what* something is a necessary condition.

Broad's description of the reasoning employed and the suppressed premisses needed, when the two methods are used for their primary task, is confined to the special case, when there are only two 'instantial' premisses and no admission of Complexity of Conditions.

Broad also formalizes a use of the Method of Agreement for finding sufficient conditions and a use of the Method of Difference for finding necessary conditions of a given phenomenon. For this purpose he has to rely on *universalized* instantial premisses, *i.e.* premisses of the form '*all ABC* is *abc*', etc. (The same form is used also in Broad's account of what I called the primary task of the two methods, but here the universalization is inessential – and therefore in my opinion misleading.) This universalization of the premises he nowhere explains.[108] It makes it possible for him to 'read off' relations of sufficient conditionship directly from the instantial premises themselves. (For example: that *ABC* is a s.c. of *abc*.) From the relations, thus established, in combination with suitable 'suppressed premises' the desired conclusions follow. The reasoning is quite correct. But I doubt whether the resulting figures can be called formalizations of Mill's canons. As far as I can see, there is no universalization of the instantial premises even tacitly presupposed in Mill's description of his methods.

Mill's description of his Joint Method is notoriously confused, and Broad has no difficulty in subjecting it to a devastating criticism.[109] Then he goes on to suggest an interesting improvement of the method. Broad's improvement makes clear the proper purpose of the Joint Method, which is to *combine* the use of a canon for eliminating possible necessary conditions with the use of a canon for eliminating possible sufficient conditions so as to obtain a method for ascertaining the necessary-and-sufficient conditions of a given phenomenon. Broad seems to think[110] that this combined method, though 'important and legitimate', is not 'absolutely conclusive' and, if I understand him rightly, in this respect different from the methods of Agreement and of Difference when correctly formalized. This view of the matter, however, is mistaken. The mistake is suggested by a peculiarity in Broad's description which, I think, ought to be corrected. The proper conclusion of the first part of the method, as described by Broad on p. 316 in [7¹], is that *a* is the only possible (simple) necessary condition of *A*. The proper conclusion of the second part again is that non-*a* is the only possible (simple) necessary condition of non-*A* or, which means the same, that *a* is the only possible (simple) sufficient condition of *A*. Taken in combination, the conclusions of both parts amount to saying that *a* is the only possible (simple) necessary-and-sufficient condition of *A*. This is just *as* certain a conclusion as any corresponding result obtained by means of the methods of Agreement and Difference. Broad's reservation to the conclusion of each part: 'strong presumption, though never a rigid proof', is therefore out of place.

6. The second part of [7] treats of Laws of Correlated Variation of Determinates. Considered as an original contribution to the subject,

this part is not nearly as important as the first part of [7]. It is essentially a formalization of W.E. Johnson's[111] 'figures of demonstrative induction'. The basic ideas are Johnson's. But technically Broad's treatment constitutes a great improvement. Not only is it much clearer but it also puts right some errors in Johnson's account. On one point Broad makes a rather severe criticism of Johnson's views (*vide infra*).

The rather formidable looking symbolism which Broad employs in this paper need not deter any reader. As a notation it is perfectly perspicuous, and the ideas expressed by its means are very simple. [...] Indeed, there may be some justification for saying that these ideas are almost trivial. Without wishing in the least to exaggerate either the intrinsic logical interest or the practical scientific importance of a correct formalization of the ideas behind Mill's methods, it seems to me undeniable that they far outweigh the interest and importance of anything that has hitherto been accomplished in the study of Laws of Correlated Variation. This study is, as a branch of inductive logic, still almost undeveloped.

Broad first states the postulates, which are needed for the purpose of constructing the figures in the form of rigid demonstrations.

The first postulate needed specifically for establishing correlation laws Broad calls the 'Postulate of the Uniqueness of the Determinate Total Effect'. Let C and E stand for determinables and c and e for arbitrary determinate values under them. Let C and E be related as total cause to total effect. (See above pp. 259–260.) Then the postulate says that, if the value c of C is *once* accompanied by the value e of E, then c of C will always be accompanied by e of E. This postulate may be said to serve the purpose of universalizing instantial premisses.

The second postulate Broad calls the 'Postulate of Variational Independence'. Let C and E again be related as total cause to total effect. Let C be a conjunction of some m and E a conjunction of some n determinables. The postulate now says that any given distribution of determinate values over the m determinables in C is a (logical and causal) possibility, and so is also any given distribution of determinate values over the n determinables in E. I hope this is a correct rendering of the thought behind Broad's formulation[112] which I do not find myself quite clear.

From this postulate, in combination with the first postulate, Broad deduces[113] a number of consequences relating to the (finite, denumerable, or non-denumerable) number of determinates under the cause- and effect-determinables. (If disjunctive determinables are allowed, Broad's conclusions would seem to require some modification.) Of most interest seems to me his discussion of the question, whether there can be an infinite number of determinates under the total cause and yet only a finite number of determinates under the total effect. The postulates are compatible with this possibility. Broad discusses[114] an example from science which looks like an actualization of it. But he also shows that, on closer examination, the case is, in the aspect under discussion, different from what it looks like. I wish someone could produce a genuine example and I should feel pretty certain that such examples exist.

The third and last postulate which is needed Broad calls the 'Postulate of Variational Relevance' (or Irrelevance, depending upon the way it is formulated). Let C and E be related as total cause to total effect. Let C be a conjunction of m determinables. Consider a distribution of determinates over the m determinables. Then consider another distribution which we get from the first by replacing the value of the determinable C_i by a different value, the values of all the other $m-1$ determinables in C remaining unchanged. The postulate now says that, if to these two distributions correspond two different determinate values of E, then to *any* two different values of C there will, the values of all the other $m-1$ determinables remaining unchanged, correspond two different values of E.

On this third postulate Broad makes the important comment that it certainly cannot be universally valid. Any law which takes the form of a *periodic* function will not satisfy it. For, it is characteristic of such laws that, when the values of the cause-factors differ by a multiple of the 'period', the value of the effect-factor repeats itself. And, as Broad points out, not only is there no *a priori* objection to such laws, but some important natural phenomena are in fact governed by laws of this kind.

Of the first postulate I should like to point out the following: Let us consider what it would be for the postulate to be false. It would mean that, *although* C is a sufficient condition of E, it may happen that to one and the same determinate value c of C there would on some occasions answer a value e_1 of E, on other occasions a value e_2, *etc*. This possibility is certainly not ruled out by the mere fact that C is a (smallest) sufficient condition of E. For this only means that to *any* occurrence of C in *any* of its determinate values c there answers an occurrence of E in *some* of *its* determinate values e. And, granted that there is this regularity, why should there then also be the further regularity to the effect that the e which answers to a given c must always be *one and the same* value e? I doubt whether any good reasons could be given for the truth of the postulate. Unfortunately I have not been able to produce any good example of a case, where the relation of C to E is that of cause to effect and yet the relation of c's to e's is not a many-to-one or a one-to-one correspondence.

Having stated the postulates, Broad restates and explains Johnson's 'Figures of Induction'. The statement of the figures is elegant and perspicuous and supersedes Johnson's own statement of them completely. I shall not dwell upon the topic here, since I find it of minor interest only. It should be noted that Broad's formalizations of these 'figures' actually answers – unlike his formal statement of Mill's methods – to the general pattern of demonstrative induction which, following Johnson, Broad sketches at the beginning of the first part of [7] and which I ventured to criticize for being too narrow. (See above p. 255.)

IV

Broad's critical notices and papers in *Mind* [2], [3], [4], [9], [10], [11] may be said to constitute a running commentary on the developments in inductive logic from the appearance of Keynes' *Treatise* in 1921 to Kneale's *Probability and Induction* in 1949. Broad has a rare talent for synoptic presentation. This makes some of his reviews excellent summaries of the reviewed works. When he makes criticisms, it is usually not in order to express a divergent opinion from the author's on some controversial issue, but in order either to point out some factual error or to suggest a clearer formulation of what he understands to be the author's intended meaning. His criticism of errors seems to me nearly always conclusive and his criticism of formulations is such as to oblige both the authors and the readers of the reviewed books to gratitude.

1. In his review [2] of Keynes' *Treatise* Broad touches upon a question which he calls 'extremely puzzling' and of which he says that he knows of no writers except himself and Keynes, who have even raised it. The

self-reference is to Broad's first published book *Perception, Physics, and Reality*. The question could be put thus: Why do we prefer probabilities relative to many data? Keynes does not answer it. Broad, however, has the following suggestion to make[115]:

> I think that our preference must be bound up in some way with the notion that to every event there is a finite set of conditions relative to which the event is certain to happen or certain not to happen. So long as the evidence is scanty a high probability with respect to it does not make it reasonable to act as if we knew that the event would happen, because it is reasonable to suppose that we have only got hold of a very small section of the total conditions and that the missing ones may be such as to be strongly relevant in an unfavourable direction. If the probability remains high relative to a nearly exhaustive set of data we feel that there is less danger that the missing data may act in the opposite direction. In fact, what we assume is that a high probability with respect to a wide set of data is a sign of certainty with respect to the *complete* set of relevant data.

This is a most interesting suggestion. It seems to me that it contains something which is possibly not right, but also something which is essentially right and points towards a solution of the puzzle.

What I doubt is the suggested connexion between preference for many data and belief in determinism. For obviously we prefer probabilities relative to many data, not only in the case of very high or very low probabilities, but also in the case of intermediate probabilities. This preference, moreover, seems to be quite independent of whether the intermediate probabilities show a tendency or not to approach extreme values (0 or 1) when the number of data is increased.

I would suggest myself that the preference in question is bound up, not with belief in determinism, but with the idea that there exists something which could be called 'complete sets of data which are relevant to the probability of a given event'. (This notion is not unproblematic, but we cannot discuss its problems here. *If* determinism, in the sense of the above quotation from Broad [2], is true, there exist, for every event, such complete sets of relevant data. Some such set, moreover, is then included in every set of conditions relative to which the event in question is certain to happen.)

If this suggestion is accepted, then our preference for probabilities based on many data can be grounded on the following second-order probability-argument: The more data we actually consider when we determine the probability of an event, the more probable does it become

265

that the set of considered data will include a complete set of relevant data. This argument, unless I am mistaken, is essentially the 'point' of Broad's suggestion, although he limits it (unnecessarily, I believe) to sets of determining conditions only.

The nature and validity of this argument demands further clarification. It cannot be given here. I believe that the argument easily can be 'formalized' within the calculus. There is no strong reason for thinking that the second-order probability in question has a different 'meaning' from the first-order probability. But there is reason to think that we cannot, in normal cases, estimate its numerical value.

The question, why we prefer probabilities relative to many data, seems to me to be one of the fundamental questions in the philosophy of probability. To have been the first to state the question clearly and to suggest an answer to it, is a noteworthy merit of Broad's.[116]

2. From Broad's very full review [9] of von Mises' *Wahrscheinlichkeit, Statistik und Wahrheit* I shall select one point for discussion. It concerns the notion of a so-called 'collective', *i.e.* a potentially infinite sequence in which the relative frequency of a certain randomly distributed characteristic is assumed to approach a limiting value. This notion seems plausible enough when regarded as an extrapolation from experiential sequences such as throws with a die or other outcomes in games of chance. But, as Broad observes[117] 'the case of vital statistics would seem to be somewhat different'. Consider, for example, the probability that a certain Mr. Smith, who is now 40, will survive his next birthday. There is an important lack of symmetry between this case and that of a die. It has to do with the fact that, to quote Broad, 'each man can die but once'. We can carry out experiments with *one and the same die* to see how often, in average, it shows, say, a 'six' and then extrapolate a limiting-frequency from the observations. But we cannot experiment with *one and the same man* of 40 to see how often, in average, he survives his next birthday. The experiment would require that we could make a man live through his 40th year any number of times. And this would be possible only if human bodies, like watches, could be 'wound up' after each lifespan.

Unless I am completely mistaken, the discovered asymmetry has far-reaching consequences for probability-theory. If the notion of a collective is to apply to the case from vital statistics at all, the collective must be identifiable with some class of 40 years old men, of which Mr. Smith is a member. Depending upon the choice of this class, the probability may be different. *This* need not be a cause of worries, since it is plausible to think that all probability is *relative* to some evidence. The difficulty is one of being able to identify the collective in question with any class at all, of which Mr. Smith is a member. For it is plausible to think that the death-rate in every potentially infinite class of human beings will be subject to *alterations* following changes in the climatic, economic, hygienic, *etc.* conditions, under which the members of the class happen to live, and there is usually no ground for thinking that the rate in question will approximate even to *some* limit. And as a consequence of this instability in the conditions determining the frequencies in such classes, the notion of a frequency-limit and therewith also von Mises' notion of a collective threatens to become *inapplicable* to them. This, I believe, is the force of Broad's observation.

Cases in which statistical observation is relevant to the determination of probabilities seem on the whole to be more like the above case from vital statistics than cases from games of chance. (In games of chance, probabilities can as a rule be hypothetically determined on the basis of *a priori* considerations, pertaining to the nature of the chance-machine itself.) It is interesting to note that von Mises' theory, which is sometimes called the *statistical* theory of probability, seems to be particularly ill-suited to deal with probabilities in statistics! The doubt which Broad raises about the applicability of the notion of a collective to vital statistics seems to me therefore to be extremely important. And it has *not* received the attention which it deserves.

3. A main theme of [8] and [11] is the character of laws of nature. Roughly speaking: Are laws 'mere' generalizations from observed uniformities or are they principles of necessitation? Speaking in particular of causation: 'Is the regularity view or the entailment view of the causal relation right? In [8] (p. 102; present volume p. 175) Broad writes:

My conclusion is as follows. Either (a) I do rationally cognize some principle which, in conjunction with suitable empirical premises, would justify me in believing certain laws of functional *entailment*, although I cannot elicit or formulate any such principle; or (b) no empirical evidence, however regular, varied, and extensive, gives me the slightest ground for believing a law even of functional *regularity*. I should tend, *prima facie*, to reject (b), as contrary to common-sense and to my own unquestioning convictions when not philosophizing about induction. But, when I realize that rejecting (b) entails accepting (a), I become more and more doubtful as to what I ought to hold.

This seems to me to be a good way of expressing the traditional dilemma in the philosophy of induction since the day of Hume. Few have seen the dilemma with such extraordinary clarity as Broad. He has, as far as I know, never been very hopeful about a way out of it. It is perhaps a fair characterization to say that he has objected with his sentiment to the idea that induction has no justification, and equally objected with his intelligence to the consolation-grounds offered by the anti-Humeans.

I believe that there is a way out of the dilemma, but I doubt whether Broad would accept it as being even a possibility. This way starts, not with the question whether there *exists* any ground for belief in induction, but with the question what we *call* a 'grounded' as opposed to a 'groundless' inductive belief and what we *mean* by 'rationally to believe' a generalization. I would suggest that, ultimately, what we call grounds of rational belief in induction are what Broad in the above quotation calls 'empirical premisses', *without* the additional support of some 'principles' (about determinism, limited variety, equi-probability, or the rest). I think, in other words, that the argumentation in the above quotation is wrong. Rejecting (b) does *not* entail accepting (a). The dilemma, as Broad sees it, simply does not exist. I shall not here do anything to vindicate my own position and beliefs in the matter. My aim with these remarks has merely been to provoke, if possible, some comments of Broad's on the problem of 'the ground of induction.' If I succeed, I know that readers of this volume will be grateful.

4. In his lecture [5] on *The Philosophy of Francis Bacon* Broad has given an appreciation of the greatest of all workers in the field of inductive logic. Besides an excellent synopsis of the plan of the Great Instauration and summary of the main content of those parts of the gigantic work which were completed, the paper contains an original and valuable attempt to clarify the meaning of 'limited independent variety' as a basic principle in the inductive philosophy of Bacon.

The supreme task of science, according to Bacon, is to find the forms of simple natures. Since the form is a necessary-and-sufficient condition of the nature, it follows that any simple nature can have only one distinct form. The number of forms and simple natures must thus be the same. Since the forms are assumed to be few in number, the number of simple natures must be few too. Broad goes on to conclude[118] that Bacon by 'simple natures' evidently means *generic* physical properties; *i.e.* determinables such as colour, temperature, density, *etc.* and *not* determinates falling under them. This seems to me logical, and would tend to make the postulated limited variety useless for the purpose of guaranteeing the certainty of most inductions.

Broad next distinguishes four different senses of possible Limited Variety in the material world. The first is that nature is composed of *kinds* of substance. This requires limited variety in the sense that each kind should be uniquely distinguished from all the rest by a comparatively small number of characteristics. (It is *this* form of limitation in nature and its relevance to induction which Broad himself had considered in [1].) If to the requirement just mentioned is added that the *number of kinds* itself should be comparatively small, we get a second form of the limited variety postulate. A third form is that the various specific modifications in a single generic property, such as colour, can be reduced to modifications in a single numerical determinable, such as frequency of lightwaves. A fourth form is that the various generic properties or simple natures reduce to specific differences of some few supreme genera. For example: colour and temperature may both be reduced to specific forms of movement.

Broad thinks[119] that, though Bacon did not distinguish these various cases of limited variety from each other, he meant to assert them all. It seems to me that the fourth case comes nearest to Bacon's actual conception of the constitution of the material world. I doubt whether he can be said to have contemplated limited variety in the first of the above senses.

Broad's paper on Bacon concludes with the often quoted question, whether we may venture to hope that, when Bacon's next centenary is celebrated, Inductive Reasoning, which has long been the glory of Science, will have ceased to be the scandal of Philosophy. I think this hope is not unfounded, in view both of recent progress in the logical study of induction and of recent investigation into the nature of the alleged 'scandal'. Progress in both branches have largely come from one and the same geographical quarter, Cambridge, – the birth-place of modern logic and modern philosophical analysis.

NOTES

Editor's note: Cross-references to Professor Broad's works as they appear in the present volume are added in these notes. They are indicated by the letters 'p.v.'.

[1] [7^I], p. 302 (p.v., p. 127).
[2] I use "Confirmation-Theory" here to mean a theory of the way in which the probability of an inductive generalization is affected by the confirmation of individual instances which fall under it. In recent years the term has acquired an inflated use and come to mean the study of the probability-relation, as such, within a so-called range-model.
[3] Cf. Broad's statement on the relation of his views on induction to those of Keynes' in [2], p. 81 (p.v., p. 63).
[4] [1^I], p. 389 (p.v., p. 1).
[5] [1^I], p. 390 (p.v., p. 2).
[6] [1^I], p. 390f (p.v., p. 2).
[7] [1^I], p. 391 (p.v., p. 3).
[8] [1^I], p. 392 (p.v., p. 4).
[9] [1^I], p. 389 (p.v., p. 1).
[10] [1^I], p. 394, b (iii) (p.v., p. 6).
[11] [1^I], p. 394, c (i) (p.v., p. 6).
[12] [1^I], p. 395 (p.v., p. 7).
[13] [1^I], p. 395 (p.v., p. 7).
[14] [1^I], p. 395 (p.v., p. 8).
[15] [1^I], p. 396 (p.v., p. 8).
[16] [1^I], p. 400–402 (p.v., pp. 13–14).
[17] *Op. cit.*, Ch. XX.
[18] [1^I], p. 402 (p.v., p. 14).
[19] [1^I], p. 400 (p.v., p. 12).
[20] [1^I], p. 389 and 404; [1^{II}], p. 42 and 45 (p.v., pp. 1, 16 and 49, 52, respectively).
[21] [1^{II}], p. 13 (p.v., p. 19).
[22] [1^{II}], p. 13 (p.v., p. 19).
[23] [1^{II}], p. 15f (p.v., p. 21f).
[24] [1^{II}], p. 16 (p.v., p. 22).
[25] [1^{II}], p. 16 (p.v., p. 22).
[26] As far as I can see Broad nowhere defines the notion of a Natural Kind. He would probably (cf. [1^{II}], p. 16; p.v., pp. 22–23) accept Mill's definition.
[27] The question is discussed in Sections 10–13 and 20 of the second part of [1].
[28] [1^{II}], p. 26 (p.v., p. 33).
[29] [1^{II}], p. 27 (p.v., p. 33).
[30] *Ibid.*
[31] *Ibid.*
[32] *Ibid.*
[33] [1^{II}], p. 27f (p.v., p. 33f).
[34] [1^{II}], p. 29 (p.v., p. 35).
[35] [1^{II}], p. 29f (p.v., p. 36).
[36] [1^{II}], p. 30 (p.v., p. 36f).

[37] [1^II], p. 43 (p.v., p. 50).
[38] Cf. the conclusion in Section 20 of the second part of [1].
[39] [1^II], p. 31 (p.v., p. 37).
[40] Yet this appears to be implicitly contradicted by the remarks on p. 44 in [1^II] (p.v., p. 51).
[41] [1^II], p. 20f (p.v., p. 26f).
[42] [1^II], p. 39 (p.v., pp. 45–46).
[43] A kind will be called 'ideal', if its instances are *exactly* similar. See [1^II], p. 38 (p.v., p. 45).
[44] [1^II], p. 37f (p.v., pp. 44–45).
[45] [1^II], p. 40 (p.v., p. 47). Italics mine.
[46] [1^II], p. 34 (p.v., p. 41).
[47] [1^II], p. 44 (p.v., p. 51).
[48] [6], p. 1f (p.v., p. 86f).
[49] [6], p. 2 (p.v., p. 86).
[50] [6], p. 7 (p.v., p. 91). See also above p. 233.
[51] [6], p. 7 and *passim* (p.v., p. 93).
[52] [6], p. 7 and *passim* (p.v., p. 93).
[53] [6], p. 8 (p.v., p. 92).
[54] [6], p. 9 (p.v., p. 93).
[55] See above p. 318 (p.v., p. 233).
[56] [6], p. 3 (p.v., p. 88).
[57] In [2] Broad discusses, in some detail, Keynes' and in [11] Kneale's version of the Principle of Indifference.
[58] *Logical Foundations of Probability*, University of Chicago Press, Chicago, 1950, § 110 A.
[59] [6], p. 12 (p.v., p. 95).
[60] [6], p. 14f (p.v., p. 97f). I hope this is a correct rendering of Broad's idea.
[61] [6], p. 17 (p.v., pp. 99–100).
[62] [6], p. 13f (p.v., p. 97).
[63] [6], p. 15 (p.v., p. 98).
[64] [6], p. 15 (p.v., p. 98).
[65] [6], p. 16 (p.v., p. 99).
[66] [6], p. 18 (p.v., p. 101).
[67] Cf. [6], 22, 36–38 and 45 (p.v., 104, 114–116 and 121).
[68] See above p. 233.
[69] [6], p. 19 (p.v., p. 101).
[70] [6], p. 22 (p.v., p. 104).
[71] Cf. [6], p. 23 (p.v., p. 104).
[72] [6], p. 25f (p.v., p. 106f).
[73] [6], p. 42 (p.v., p. 119).
[74] See [6], 30f (p.v. 110), and above p. 251.
[75] Cf. Broad's definition of a *generalization* in [6], p. 25 (p.v., p. 106).
[76] [6], p. 25 (p.v., p. 106).
[77] Cf. [6], p. 31 (p.v., p. 111).
[78] [6], p. 27 (p.v., p. 108).
[79] [6], p. 31 (p.v., p. 111).
[80] [6], p. 30 (p.v., p. 110).
[81] [6], p. 27–30 (p.v., pp. 108–110).

INDUCTION, PROBABILITY, AND CAUSATION

[82] [6], p. 34–36 (p.v., pp. 113–116).
[83] [6], p. 38–41 (p.v., pp. 116–118).
[84] [6], p. 39 (p.v., p. 117).
[85] [6], p. 41f (p.v., p. 118f).
[86] [6], p. 41 (p.v., p. 118).
[87] [6], p. 41f (p.v., p. 119).
[88] [6], p. 42 (p.v., p. 119).
[89] [6], p. 43 (p.v., p. 120).
[90] [6], p. 45 (p.v., p. 121).
[91] [6], p. 43 (p.v., p. 120).
[92] [6], p. 43f (p.v., p. 120).
[93] [7^I], p. 302ff (p.v., p. 127ff).
[94] Elements of Logic, Bk. IV, Ch. I, § 1.
[95] [7^I], p. 304 (p.v., p. 130).
[96] Sometimes the relation of 'cause' to 'condition' is even more complex. Thus, e.g., in the classical theory of inverse probability, which is also traditionally said to be a theory for estimating the probability of causes, 'cause' (usually) means a member of a disjunction which is a necessary condition of a certain phenomenon. For such causes Broad (in [10], p. 16, cf. p.v pp. 273–276) has introduced the term *substitutable requirements*. See also above pp. 259–260.
[97] [7^I], p. 306 (p.v., p. 131).
[98] [7^I], p. 311 (p.v., p. 137).
[99] [7^I], p. 310 (p.v., p. 136).
[100] [7^I], p. 310 (p.v., p. 136).
[101] [7^I], p. 305 (p.v., p. 131).
[102] [7^I], p. 307 (p.v., p. 132).
[103] [7^I], p. 308, Proposition (6) (p.v., p. 134).
[103a] *Editor's note*: A restatement and further development of some of them is found in the present volume, pp. 273–276.
[104] [10], p. 17 (cf. p.v pp. 273–276, especially p. 274).
[105] [7^I], p. 311f (p.v., p. 138).
[106] [7^I], p. 312 (p.v., p. 139).
[107] [7^I], p. 313 (p.v., p. 140).
[108] I think the reader of Broad's paper will have to consult Johnson's *Logic*, Part II, Ch. X, especially § 4 and § 5 for an explanation.
[109] [7^I], 315f (p.v., p. 141f).
[110] [7^I], 317 (p.v., p. 143).
[111] *Logic,* Part II, Ch. X, §§ 8–20.
[112] [7^{II}], p. 427 (p.v., p. 145).
[113] [7^{II}], p. 427–431 (p.v., pp. 145–149).
[114] [7^{II}], p. 429f (p.v., p. 147f).
[115] [2], p. 78 (p.v., p. 60).
[116] For a discussion of Broad's question see J. Hosiasson, 'Why do we prefer probabilities relative to many data?', *Mind* **40** (1931).
[117] [9], p. 480 (p.v., p. 186).
[118] [5], p. 131. Quoted from *Ethics and the History of Philosophy*, Routledge and Kegan Paul, London, 1952.

C. D. BROAD

REPLIES TO MY CRITICS

ANALYTICAL QUESTIONS

Professor Russell very justifiably finds much that is obscure in my remarks about causation in *Examination of McTaggart's Philosophy* Vol. I, Chapter XIII. He tries to lighten the darkness by restating what I may have had in mind in terms of the notions of necessary condition and sufficient condition. I am sure that that is the right course. I have pursued it myself in later writings. I think that the simplest way for me to clear up the matter is to begin by giving some definitions and making some statements based on the contents of pp. 15 to 18 of the first of my papers entitled, 'Hr. von Wright on the Logic of Induction I–III', in *Mind* (1944).

(i) P is a *sufficient precursor* ('S.Pr.') of Q, if from any instant into which a P-event were to enter a Q-event would issue.

(ii) P is a *necessary precursor* ('N.Pr.') of Q, if into any instant from which a Q-event were to issue a P-event would have entered.

(iii) P is a *smallest sufficient precursor* ('S.S.Pr.') of Q, if (a) it is a sufficient precursor of Q, and (b) it is either (α) a simple characteristic p, or (β) a conjunctive characteristic $p_1 \& p_2 \& ... p_n$, such that, if any of the conjuncts be omitted, what remains is *not* a sufficient precursor of Q.

(iv) A *contributory precursive condition* ('Cy.Pr.Cn.') of Q in any simple characteristic, or any conjunction of such characteristics, which is a conjunct in a S.S.Pr. of Q.

(v) P is a *smallest necessary precursor* ('S.N.Pr.') of Q, if (a) it is a necessary precursor of Q, and (b) it is either (α) a simple characteristic p, or (β) a *disjunctive* characteristic p_1 or-p_2 or-...p_n, such that, if any of the alternants be omitted, what remains is *not* a necessary precursor of Q.

(vi) A *substitutable precursive requirement* ('Sb.Pr.Rq.') of Q is any simple characteristic, or any disjunction of such characteristics, which is an alternant in any S.N.Pr. of Q.

(vii) If Q has *only one* S.S.Pr., every conjunct in it may be described as an *indispensable contributory precursive condition* ('I.Cy.Pr.Cn.') of Q. If, on the other hand, Q has several alternative S.S.Pr's, then any characteristic which is a *conjunct in all of them* may be so described.

So much by way of definition. It is important to be clear about the logical relationships of the two notions of *necessary precursor* and *indispensable contributory precursive condition*.

(a) It is logically possible for P to be an I.Cy.Pr.Cn. of Q *without being* a N.Pr. of Q. For, whether there be only one S.S.Pr. of Q or several alternative S.S.Pr's of Q, it remains *logically possible* that there should be cases in which a Q-event issues from an instant into which *no* S.S.Pr. of Q has entered. Now, in order for P to be a N.Pr. of Q, a P-event would have to enter into *every* instant from which a Q-event issues. So a P-event would have to enter *inter alia* into those instants (if any) from which a Q-event issues *without any* S.S.Pr. of Q having entered. But, in order for P to be an I.Cy.Pr.Cn. of Q, it has only to be a conjunct in *every S.S.Pr.* of Q. Obviously that does not guarantee the entry of a P-event into those instants from which a Q-event issues without any S.S.Pr. of Q having entered. Since it is logically possible that there should be such instants, it is logically possible for P to be an I.Cy.Pr.Cn. of Q without being a N.Pr. of Q.

(b) This possibility would be ruled out, if and only if we were to assume that in *every* case in which a Q-event issues from an instant there is *some* S.S.Pr. of Q. This might be called the *Postulate of Smallest Sufficient Precursors*. On that assumption any I.Cy.Pr.Cn. of Q must be a N.Pr. of Q.

(c) It is logically necessary that a conjunction of all the I.Cy.Pr.Cn's of Q should be a S.Pr. of Q. But it is *not* logically necessary that a conjunction of all the N.Pr's of Q should be a S.Pr. of Q. The latter proposition would, however, follow from the Postulate of Smallest Sufficient Precursors.

Now it is certain that neither the above distinctions, nor consequently the logical relations between the notions distinguished, were clearly before my mind when I wrote the chapter on Causation. In terms of them I will make the following comments on certain things which I wrote there.

(i) I think that what is generally understood by the phrase '*total cause* of such and such a change issuing from an instant *t'* is a S.S.Pr. for changes of that kind. Therefore the most obvious interpretation of the sentence: 'All changes of such and such a kind are caused' would be that in every case where a change of that kind issues from an instant there is a S.S.Pr. for it. That would be quite consistent with holding that there is a *plurality* of alternative S.S.Pr.'s for changes of that kind; that in some cases one is present and in other cases another; and that perhaps in some cases several of them are present together. The most obvious interpretation of the sentence: '*All* changes *whatever* are caused' would be a generalization of the above statement about all changes of *such and such a kind*. It would in fact be the Postulate of Smallest Sufficient Precursors.

(ii) Any reader might be excused for thinking that it was *this* proposition which I claimed to find self-evident when I wrote (*Examination* Vol. I, p. 232) 'Every change has a cause', and said that this was to me evidently true. But in fact I did *not*, and do *not*, find it self-evident that for every case in which a change of any kind issues from an instant there must be a S.S.Pr. for a change of that kind issuing from that instant.

If the reader should continue until he reaches the discussion of voluntary decision on p. 238 of the volume in question, he will find that what I there claim to be self-evident would be expressed (at any rate to a first approximation) by the following sentence: 'In every case in which a change of any kind issues from an instant there must be a change entering into that instant, such that a change of the former kind *would not have* issued *unless* one of the latter kind had entered'. Now this, as Professor Russell rightly points out, is an assertion about *necessary* precursors, and not about sufficient precursors.

The above statement needs a certain amount of elucidatory comment, in view of the fact (which I did not recognise at the time) that a N.Pr. need not be simple, and that the S.N.Pr. for a given kind of change may be a *disjunction*. (I owe the recognition of this to Professor von Wright.)

It might be that an event of the Q-kind would not issue from any instant unless an event of a certain kind P_1-or-P_2 should have entered into that instant, but that in some cases the entering event is of the kind P_1 and there is none of the kind P_2, that in others it is of the kind

P_2 and there is none of the kind P_1, and that in yet others perhaps there is either a single entering event of the two kinds or two entering events one of each kind.

What I claimed to find self-evident might therefore be re-stated as follows. The issuing of an event of any given kind (say Q) from any instant must be preceded by the entry into that instant of an event which is either (a) of a certain *one* kind (the same in all such cases), or (b) of one or another of a *certain limited number* of alternative kinds (in some such cases of one, and in other such cases of another, of these alternative kinds.)

I think I may say of this proposition the following two things. (a) The contradictory of it is certainty not self-contradictory. (b) When I reflect on the contradictory of it, and try to consider 'what such a state of affairs would be like', I find it almost impossible to think that it could be true.

(iii) I should not now be inclined to attach much, if any, weight to the proposition which I asserted, at the bottom of p. 233 of Vol. I of *Examination*, to be self-evident. This to the effect that a given change issuing from a given instant cannot have 'more than one total cause'. I should now identify '*a* total cause' of a particular change with any S.S.Pr. of such changes which enters into the instant from which that change issues. If there should be only one S.S.Pr. entering into the instant in question, we can talk of '*the* total cause' of the change on that occasion. But such a change may have several alternative S.S.Pr's, and it does not seem to me self-evidently impossible that more than one of them should enter into a given instant from which such a change issues. In that case, it seems to me, we must be content to say that the particular change in question has *several coexisting* total causes, and therefore that there is nothing that can be called '*the* total cause' of it. I should describe such a change as 'over-determined'. It would be easy to produce quite plausible *prima facie* instances of over-determination.

INDUCTION AND LAWS OF NATURE

It will be convenient to discuss this topic immediately after the above discussion of the notion of Causation. It forms the subject of essays by Dr. Hanson, Professor Nelson, and Professor von Wright.[2] [...]

Induction

The topics treated in Professor Nelson's and Professor von Wright's essays partly overlap and partly diverge, so in some of the sub-sections which follow I shall be concerned with what is common to both and in others with what is peculiar to one or the other.

(1) *The so-called 'Problem of Induction'.* This question is treated by both writers. Professor von Wright quotes a dilemma, in which I summed up my position in the paper entitled 'Mechanical and Teleological Causation', in Aristotelian Society's *Supplementary Volume XIV*.[3] He suggests that, instead of pursuing the course which seems inevitably to end in that dilemma, we should begin with the question: What do we *mean* by calling an inductive belief 'grounded' (as opposed to 'groundless' or 'ill-founded'), and what do we *mean* by 'rationally believing' in reference to an inductive generalisation? He is inclined to think that, if we do this, we shall see that what we call 'grounds of rational belief in induction' are *just* empirical premisses *without* support of any general principles. He does not attempt to argue his case, but hopes that I may comment on it. So I will take this as the text of my sermon in this sub-section.

I would suggest that what must presumably have happened in the case of *deductive* logic may be useful as an analogy and a contrast. Here, I suppose, we could distinguish in theory three stages, though very likely they overlapped historically.

(1) There would have been a number of particular bits of deductive argument which all or most sane persons *accepted* in the law-courts, in monetary calculations, in mensuration, and so on, except when under the influence of some strong desire or emotion which was known to distort a person's judgment. There would have been a number of particular bits of deductive argument which all or most sane persons, with similar qualifications, *rejected*. Finally, there would be a number of particular bits of deductive argument which were accepted by some and rejected by other sane persons when in an emotionally calm state.

(2) It would be natural, then, to compare and contrast the generally accepted with the generally rejected arguments, in order to see whether there were other features, beside general acceptance, common and peculiar to the former. This stage might be illustrated by the discovery

and formulation of the traditional rules of the syllogism. At this stage it might be agreed to be a fair test, in the case of a disputed argument, to note whether it did or did not have the characteristics which had been found to be in fact common and peculiar to arguments commonly accepted by sane men in their calmer moments.

(3) One might still, however, see *no reason why* an argument having all the characteristics in question should be valid, and *why* one which lacked any of them should be invalid. There is nothing, e.g., *obviously* wrong with a syllogism having a negative conclusion and two affirmative premises. The next stage, then, would be to try to get behind the empirical tests, and to show that they are consequences of more fundamental principles which are *self-evident*. That can be done in various alternative ways, which I need not describe here, for the rules of the syllogism.

Let us now compare and contrast this with the case of inductive arguments. In the case of *deductive* inference we are all, I suppose, agreed as to what we *mean* by calling an argument 'valid'. At any rate there is one condition which would generally be acknowledged to be necessary and sufficient for the validity of a *deductive* argument. It is this. It must be *impossible* that the premisses should be true and the conclusion false; and this impossibility must rest, not on the impossibility of the premisses (though they *may* be impossible, as in a *reductio ad absurdum* argument in pure mathematics), nor on the necessity of the conclusion (though it *may* be necessary, as it always is in the case of any valid argument from true premisses in pure mathematics), but on a certain relationship between the *logical form* of the premisses and the *logical form* of the conclusion. The task for philosophers of *deduction* is to classify arguments which answer to this admittedly necessary and sufficient condition of validity; to elicit the formal features common and peculiar to them; and then, if possible, to bring them under one or a few general principles, which all or nearly all sane and competent persons find self-evident. That programme has in the main been accomplished.

But it is not obvious what we *mean* by calling an *inductive* argument 'valid'; or, if you prefer it, there is no one condition which is generally acknowledged to be necessary and sufficient for the validity of such an argument. What is quite certain it this. If we use the accepted definition or criterion of 'validity' as applied to a *deductive* argument, and if we take the *complete* premiss of an *inductive* argument to be: *This, that, and the other S (which are all that have so far been observed) have been P*, and the conclusion to be: *All S's, past, present or future, respectively have been, are, or will be P*, then *all* inductive arguments are *invalid*. Now we all use inductive arguments, and we all accept the conclusions of many of them and guide important actions by reference to these. So reflective persons cannot but find this situation intellectually disturbing.

Now at this point there seem to be two alternatives open to us. One is to suppose that the definition or accepted necessary and sufficient condition of 'validity', as applied to *deductive* arguments, applies also to *inductive* ones. The other is to deny this, and to set out from that point. I will now say something about each of these alternatives in turn.

(1) If we are going to use the old definition or accepted necessary and sufficient condition of 'validity', and yet to admit the possibility that some inductive arguments are valid, we must try to save the situation in one or other or a combination of the following ways. We might suppose either (i) that a valid inductive argument has an additional *implicit premiss* beside the instantial propositions which are its only explicit premisses, or (ii) that the *conclusion* of a valid inductive argument must take a *weaker form* than the unqualified *All S is P*. I think that it is now quite plain that anything on these lines needs *both* expedients in order to be at all hopeful, *viz.*, adding some kind of universal premiss to the explicit instantial premisses, and stating the conclusion in terms of probability.

If that were done, it is evident that the *principles* (as distinct from the premisses) of inductive inference would include, beside those of non-problematic deductive inference, at least the formal principles of the calculus of probability, *e.g.*, the axiom of addition concerning the probability of a disjunctive proposition, and the axiom of multiplication concerning that of a conjunctive one. I do not think that this in itself would be felt to raise any special difficulty.

Anyone who follows this line will have to deal with the following three questions, which might be described respectively as 'logical', 'ontological', and 'epistemological'. (i) What are the minimal universal premisses which, if added to the explicit instantial premisses, would make very highly probable the conclusions to those inductive arguments which are commonly accepted as practically certain by sane and instructed persons? (ii) What account of the structure of nature as a whole, or of certain departments of nature, would best fit in with the assumed truth of these universal premisses? (iii) How, if at all, do we know that these premisses are true or that they are highly probable? I think that this agrees almost exactly with the scheme outlined by Professor Nelson.

Before considering the other alternative, suggested by Professor von Wright, I will make the following comments on the alternative outlined above.

(a) There is no guarantee that the whole enterprise may not break down at the first stage. In that case we should have to admit that, so far as we can tell up to date, *no* inductive arguments are valid, in the sense of 'validity' supposed, even when their conclusions are stated in terms of probability.

(b) Even if the logical problem can be solved satisfactorily, the epistemological problem would (as both Professor Nelson and I have emphasised) remain very troublesome. The additional premiss (and still more obviously the propositions about the 'structure of nature' which have to be assumed in order that it shall be applicable) must be *general*, and it cannot be *merely analytic*. Yet our acceptance of it cannot, without circularty, be based on *induction*; and, even if the possibility of necessary synthetic propositions were admitted (which it is not by most contemporary English and American philosophers), no additional premiss which has been plausibly alleged to fulfill the conditions has any trace of *self-evidence*.

(c) This leads me to the following two reflexions, (α) Even if the epistemological difficulties should be insoluble, that would not diminish the value of the analytic and the ontological sections of this line of thought. The justification of induction, where it is thought to be justifiable, would have to be stated *conditionally*, and not categorically. But even that would be no small gain in insight. (β) The situation would be remarkably like that which Kant (as I understand him) contemplated in regard to such allegedly synthetic *a priori* propositions as he held to be capable of 'transcendental proof'. People claim to *know*, or to have *good grounds for very strongly believing* certain general propositions as a result of inductive reasoning. Suppose we grant their claim. Suppose we can show that it can be valid, *if and only if* certain propositions about the structure of nature are true. Then we are entitled to accept those propositions, even though they be synthetic and though they have no trace of self-evidence. They would be 'synthetic *a priori* propositions' in precisely the sense in which Kant held that, *e.g.*, the law of universal causation and the conservation of mass are so.

(2) Let us now consider the other alternative, suggested by Professor von Wright, which is nowadays much the more popular of the two. The contention is that if an *inductive* argument can properly be described as 'valid' or 'invalid', those words must be understood in a special sense, appropriate to such arguments. On that supposition, it is of course quite possible that certain inductive arguments may be 'valid', in the appropriate sense, *without* the addition of any implicit general premiss to their explicit instantial premisses, and perhaps without reformulating their conclusions in terms of probability. On this suggestion I would make the following comments.

(i) Plainly the first task would be to formulate a definition, or generally acceptable necessary and sufficient condition, of what I will call 'inductive validity'. Here we may compare and contrast this enquiry with Stage (1) of what I supposed above to have happened in the case of *deductive* arguments. We should have to consider typical *inductive* arguments, which all or most sane persons in their calmer moments accept, and compare and contrast them with typical inductive arguments which all or most of such persons under such conditions reject. But the difference would be this. In the case of *deductive* arguments there was from the outset no doubt as to what is *meant* by 'valid' and 'invalid' as applied to them. The object of the comparison and the contrast was not to elicit the *meaning* of 'validity', but to discover, and if possible to rationalise, *tests* for its presence or absence in any deductive argument. But in the case of *inductive* arguments the primary object of this comparison and contrast would be to discover what competent persons, who use and criticise such arguments, *mean* when they call some of them 'valid' and others 'invalid'.

Unless it turned out that inductive validity had some fairly close and important analogies to deductive validity, it would be better not to use the word 'validity' or 'invalidity' of inductive arguments, but to coin some other technical term. I should think that the irreducible minimum of analogy would be that the 'validity' of an inductive argument should depend in some assignable way on relationships of *logical form* between its premises and its conclusion.

(ii) However that may be, it might still be worth while, after having elicited an agreed definition of 'inductive validity' in this or in some other way, to proceed thenceforth as logicians did with deductive arguments. That procedure would be as follows. (a) To try to discover features, other than those which enter into the definition of 'inductive validity', which are common and peculiar to arguments which are inductively valid. (b) If that can be done, to try to show *why* the presence of all these features entails inductive validity, and the absence of any of them entails inductive invalidity. (c) To try to reduce these features as far as possible to one or a few very general headings. If all this could be accomplished, there would remain the following typically philosophical questions. What is the nature of the ultimate principles on which the tests for inductive validity rest? Are all of them analytic, or are some of them synthetic? If some of them are synthetic, how are they known or rationally believed to be true?

(iii) Now it *might* happen that, when one elicited the meaning of 'inductive validity', the consequence which Professor von Wright thinks would follow, *viz.*, that the grounds of rational belief in induction are *just* empirical premises *without* support of any general principles, would be seen to follow. Or it might not. All that I will say in conclusion is this. We must of course distinguish between the *premises* of a valid argument, and the *principles* which the argument exemplifies and which ensure and make evident its validity. In the valid syllogism, *e.g.*, *All men are mortal, and all Greeks are men, therefore all Greeks are mortal* the only *premises* are the two propositions which are stated before the word 'therefore'. The *principles* which the argument exemplifies, and which together make evident its validity, are such propositions as the following: (a) If a class is empty, every sub-class of it is empty; and (b) If every member of an exhaustive set of sub-classes of a class is empty, then that class is empty. Now I should think it certain that, if there are any *principles* for the 'validity' of inductive arguments (no matter what meaning be attached to 'inductive validity'), they must be *general* propositions. But that would leave open the question whether a valid inductive argument does or does not have to include one or more general propositions among its *premises*.

REPLIES TO MY CRITICS

(2) *Professor Nelson's account of Inductive Argument.* I found this of very great interest. I will first try to state it, as I understand it, in my own words, and will then make a few comments on it. To simplify the exposition I will confine myself to inductive generalization where the instantial premiss is that N instances of S have been observed (say $S_1, S_2, ..., S_N$) and that all of them have been P. With that understanding I would summarise Professor Nelson's theory as follows.

(i) If the argument is to be defensible, the conclusion must not take the unqualified non-modal form *All S is P*. It must take the form *It is likely, to such and such a degree, that all S is P*.

(ii) This must be carefully distinguished from any statement of the form: 'The proposition *All S is P* has such and such a degree of probability with respect to the datum q.' The following points are very important to notice here. (a) 'Likely', in the sense in which Professor Nelson uses it, is analogous (except in that what it stands for is capable of degree) to 'true'. (b) On the other hand, the statement that p has such and such a degree of probability with respect to q is comparable to the statement that p is entailed by q. Like it, it is a statement which is *necessarily* true or *necessarily* false, as the case may be. And, like it, its truth or falsity depends on certain relations between the *forms* of p and of q, and not on their individual necessity or impossibility, truth or falsity, likelihood or unlikelihood.

(iii) Nevertheless, in order to establish inductively the conclusion *It is likely to such and such a degree that all S is P*, we require such a proposition as is expressed by the sentence: 'With respect to the proposition that N instances of S have been observed and all of them have been P, it is probable to such and such a degree that all S is P'. I will symbolise the proposition, expressed by the sentence in inverted commas, by $\Pi_N(\sigma)$ where σ is the degree of probability in question.

(iv) The part played by $\Pi_N(\sigma)$ in establishing a conclusion *inductively* may be compared with that which is played, in establishing *deductively* that all Greeks are mortal from the premiss that all men are mortal and all Greeks are men, by the proposition which is expressed by the sentence: '*All men are mortal & All Greeks are men* entails *All Greeks are mortal*'. The following points are important to notice here:

(a) In the deductive argument we use a principle of 'Deductive Detachment'. Knowing that in fact all men *are* mortal and all Greeks *are*

283

men, we are entitled to drop those premisses and to accept as *true* the proposition that all Greeks are mortal. In the inductive argument we need a comparable principle of 'Inductive Detachment'. Knowing that in fact N instances of S *have* been examined and that all of them *were P*, we are entitled to drop that premiss and to accept as *likely to such and such a degree* that all S is P.

(b) According to Professor Nelson, the degree of likelihood which it is justifiable to assign to *All S is P*, under the conditions supposed, is a function of the degree of probability σ, which *All S is P* has with respect to the premisses in the complex proposition $\Pi_N(\sigma)$. As to this function he will say no more than the following. The degree of likelihood of *All S is P*, given that the premisses in $\Pi_N(\sigma)$ are known to be true and can therefore be dropped, increases with σ, the degree of probability of *All S is P* in respect to those premisses.

(v) The last point in the theory is this. Professor Nelson holds that we never have any good reason to accept such a proposition as $\Pi_N(\sigma)$ *on its own merits,* as we have, *e.g.*, to accept the proposition expressed by the sentence: '*All Greeks are mortal* is entailed by *All men are mortal & All Greeks are men*'. The only ground for accepting such a proposition as $\Pi_N(\sigma)$ is that it is *entailed by* a certain other proposition, which he calls the 'Principle of Induction', and that we *know* this to be *true*. We will denote this principle by P_I. It is important to note the following points about it. (a) Professor Nelson does not claim to be able to formulate it satisfactorily. But he thinks that progress has been made towards doing so, and that this is illustrated, *e.g.*, by the substitution of Keynes's 'Principle of Limited Variety' for Mill's 'Uniformity of Nature'. (b) He draws a distinction between P_I itself, and the characteristics which we must ascribe to the actual world if P_I is to be true and applicable to natural phenomena. The proposition that nature has these characteristics is ontological, whilst P_I itself is described as 'formal'.

Supposing this to be a fair account of Professor Nelson's very interesting theory, I will make the following comments.

(i) I wonder why he uses P_I as a premiss which entails $\Pi_N(\sigma)$, instead of modifying $\Pi_N(\sigma)$ by introducing P_I into it as an additional premiss. The modified proposition, which we will denote by $\Pi'_N(\sigma)$, would then be expressed by the sentence: 'The proposition *All S is P* has probability of degree σ with respect to the conjunction of P_I with the premisses of

$II_N(\sigma)$'. I do not see any obvious objection to this. And, unless there be some objection, I should think it would have one obvious advantage. For $II'_N(\sigma)$ would hold in virtue of the *form* of its premisses and its conclusion, just as a valid syllogism does; whilst $II_N(\sigma)$ would not do so (if I understand Professor Nelson aright), though I suppose that the proposition that P_I entails $II_N(\sigma)$ would do so.

(ii) Professor Nelson puts the argument in terms of a definite degree of probability σ, and a definite degree of likelihood, which increases with σ. I take it that he does not suppose that these can be exactly measured in any particular case. It would be enough that in favourable cases one should know that σ was high enough to ensure that the degree of likelihood of *All S is P* is considerable.

(iii) As regards P_I itself I have two remarks to make. (a) Taking it as a 'formal' principle, I feel rather uncomfortable about a premiss which it is admitted that no one has so far managed to formulate satisfactorily. In order to 'detach' P_I in Professor Nelson's form of the argument, one must *know* that it is true (or at any rate 'highly likely'). But unless one knows what it is, how can one know this about it? I suppose we should have to say that what one knows is that there is *some* formulable proposition (never as yet satisfactorily formulated), which has the logical properties ascribed to P_I and which is true or highly likely.

(b) I think that the distinction between the 'formal' principle and its ontological ground might be rather difficult to define. Would it come to this? The *formal* principle would state in extremely abstract terms the conditions which must be fulfilled in *any possible world* in which inductive generalization would be a valid process leading in favourable cases to highly likely conclusions. The *ontological* principle would be a much more concrete statement as to the structure of the *actual world* which ensures that these conditions are fulfilled in it.

(iv) On the notion of 'likelihood' I will make the following comments:

(a) When a person accepts a proposition (rightly or wrongly, reasonably or unreasonably) as *true*, he is prepared (so far as he is not hindered by temperamental or occasional defects, intellectual or moral) to apply it without hesitation in practice where it is relevant, to accept without question in theory any consequences which seem to him to follow from it, to use it unhesitatingly as a basis for his further reflexions and investigations, and so on. Now there is undoubtedly an attitude which we often have towards a proposition, where all this holds good with the substitution of 'with very considerable confidence' for 'unhesitatingly'. The latter may fairly be described as accepting a proposition (justifiably or unjustifiably) as *more or less likely*.

(b) One important way in which a person comes to accept a proposition as *true* is by noting that it seems to him to be logically entailed by certain other propositions, which he accepts as true. In such cases we may say that he accepts it as 'deductively established'. One important way in which a person comes to accept a general proposition as *more or less likely* is by what he takes to be a valid inductive argument from premisses which he accepts as true. These always include *at least* a number of favourably instantial propositions, together with a proposition to the effect that these are all the relevant instances that have been observed. In such cases we may say that he accepts it as 'inductively supported'.

(c) If a person accepts a proposition as *true*, because *deductively established*, he cannot hope to strengthen his case through the possible discovery of additional true propositions which entail that conclusion. These will only provide him with *alternative* lines of proof, all of which could be dispensed with, and each of which could be substituted for his original line of proof. They are like a lot of ropes, each attached to a different hook, and each amply sufficient to support a certain weight. But suppose a person accepts a general proposition as *likely to at least a certain degree*, because *inductively supported*. Then he can hope to strengthen his case (though he must also fear its complete collapse) by the examination of further relevant instances. The mere addition of further true premisses of the same kind (provided that the proposition that they include all the observed instances remains true) will inductively support the conclusion *still more strongly* and will justify one in accepting it as *likely to a still higher degree*. Here the additional true premisses are comparable to additional strands in a single rope, which is always liable suddenly to give way.

(v) Lastly, I would like to say how fully I agree with the following contention of Professor Nelson's. It is hopeless to consider the principles of induction in isolation from the other principles and categories which are involved in the notion of a world of persistent things with varying states, co-existing and inter-acting in a single spatio-temporal system.

Whatever defects there may be in Kant's discussion of the 'Principles of Pure Understanding', he had at least grasped this essential point, which his predecessors had failed to note and which most of his successors seem to have forgotten.

(3) *Assumptions about Antecedent Probability*. Professor von Wright discusses this in connexion with problems in probability concerned with drawing counters from a bag, noting their colours, and thence arguing to the probability of various propositions about the colours of the counters in the bag. I have considered such problems in 'Induction and Probability' and in 'The Principles of Problematic Induction'.[4]

In the former I assumed that the $n+1$ alternatives, that a bag containing n counters should contain 0 or 1 or...n counters of an assigned colour (e.g., white), would be equi-probable antecedently to any of them being drawn and looked at. In the latter paper, after having read Keynes's *Treatise on Probability*, I argued that this assumption leads to a contradiction. I there assumed instead that there are v distinguishable colours (including black and white), and that it is antecedently equi-probable with regard to any counter in the bag that it would have any one of these colours. Professor von Wright mentions a third possible assumption, which I did not consider in either paper, *viz.*, that every possible 'constitution' of the contents of the bag with respect to an assigned colour (e.g., white) is antecedently equally probable. He expresses regret that I did not work out the consequences of this.

Now I think that this third possible assumption can be dismissed quite briefly. I take it to be equivalent to assuming that it is antecedently equi-probable with regard to any counter in the bag that it would either *have* or *not have* the assigned colour (*e.g.*, white). For all purposes of mathematical deduction that is equivalent to putting $v=2$ in the calculations in 'Principles of Problematic Induction'. It seems to me obvious that the assumption as to equi-probability which I made there is more defensible than the assumption of equi-probability of 'constitutions'. For the latter lumps together under the heading 'other-than-white' all the remaining colours, and then counts this *disjunction of colours* as precisely on a level with the *single colour* white.

Professor von Wright says that he thinks there is no possibility of proving or of disproving any of these alternative assumptions about equi-probability. I am inclined to agree with him as to the impossibility of *proving* any of them without making factual assumptions. I think, *e.g.*, that the assumption which I made in *P. of P.I.* would be reasonable only if one had the following information, or something formally equivalent to it, *viz.*, that the bag had been filled by drawing n counters from *another* bag, which contained *equal large numbers* of counters of each of the v colours, well mixed with each other. But I should have thought that it was possible to *refute* some assumptions by showing that they lead to consequences which are plainly absurd. I do not see anything wrong with the argument by which I tried to show in *P. of P.I.* that the assumption made by me in 'Induction and Probability' leads to absurdities, if we admit that there is *more than one colour* (*e.g.*, red and blue) besides the assigned one (*e.g.*, white), which might belong to one or more of the counters in the bag.

(4) *The notion of 'Loading'*. From problems concerned with drawing counters from bags the transition is natural to problems concerned with throwing dice, spinning roulette-wheels, and so on. The notion of 'loading' has its most obvious applications in reference to the latter problems. It is discussed both by Professor von Wright and by Professor Nelson. I will take their remarks in turn.

In *P. of P.I.* I made the following assertions. (i) 'The notion of loading is the notion of a constant cause-factor which operates throughout the whole series of throws, and co-operates with other and variable cause-factors to determine the actual result of each throw.' (ii) 'I shall say that the counter is loaded to degree s in favour of red, if and only if the antecedent probability of its turning up red would be s for anyone who knew in detail how it was constructed.' Professor von Wright finds this obscure. He says that he would understand by 'load' a certain *antecedent probability*. And he asks whether I suppose the 'constant cause-factor' to be this probability *itself* or some feature in the *physical world* which may be held *responsible for* the 'load', in his sense.

The answer is that I meant the following. I thought of the load, *not* as a probability but as a *physical* factor (*e.g.*, the location of the centre of gravity at such and such a position in relation to the geometrical centre of the body in question) *determining* the antecedent probability of a face of such and such a colour coming up. It would strike me as linguistically barbarous to talk of a *probability* as a cause-factor, and I should not wittingly do so.

My statement that induction, in such cases, presupposes a reference to causation was therefore intended to mean something different from the minimum which Professor von Wright suggests that I might have meant by it. In the context it was intended to mean something like the following. The fact that the antecedent probability of a loaded die turning up a 6 on any occasion is so-and-so is determined jointly by the following facts. (i) That the position at which it comes to rest on any occasion is causally determined jointly by (a) the position of its centre of gravity in relation to its geometrical centre, (b) its geometrical, elastic, and other permanent properties, (c) the correlative properties of the surface on which it falls, and (d) the angle at which it hits the surface. (ii) That it is antecedently equally likely to hit the surface at any one of the innumerable alternative geometrically possible angles. (I suspect that this second statement would need some modification, but I think that the notion of the equi-probability of certain alternative geometrical possibilities being fulfilled would still enter.)

Passing now to Professor Nelson's 'roulette-wheel', I would make the following comments:

(i) He contrasts the case of a wheel which is 'honest' and one which is not. But ought we not rather to contrast one that is *known* to the player to be honest, and one which is *not known* to him to be so or not to be so. In the latter case the possibility that it is biased is admitted from the beginning.

If the wheel is *known* to the player to be honest, then no run of a single number, however long, and no sequence of numbers, however often repeated, would give him any rational ground for betting in favour of a repetition of that number or of that sequence. That is almost, if not quite, an analytical proposition. But, if the bare possibility of bias is admitted from the first, then it might be argued that a sufficiently preponderant proportion of a certain number, or of a certain sequence of numbers, would provide a ground for a rational belief that it *is* biased *in a certain way*. That in turn would provide a reasonable ground for betting in a corresponding way on its future behaviour.

Professor Nelson does in fact consider this kind of argument in connexion with his criticism of the 'Precept Theory'. The essential point seems to me to be one which he himself makes. A glance at the formula for the application of the principles of inverse probability shows that all that an accumulation of uniformly favourable instances can do for a hypothesis is continually to *multiply by a new factor its initial probability*. Now, if that is to lead to a final probability whose upper limit is 1, we must have reason to believe beforehand, not merely that the initial probability is greater than 0, but that the *lower limit of possible values* is greater than 0. Now that is not secured merely by the negative fact that it is not impossible that the wheel may be loaded in one way or another.

(ii) About artificial cases, such as roulette-wheels, the following points may be worth making:

(a) No one in practice is in a position to *know* (even in the popular sense of that word) that a roulette-wheel is honest. At most he may have extremely good reasons to believe that it has been made by a competent and reliable firm in accordance with the accepted method for making honest roulette-wheels, that it has not become worn or tampered with, and so on.

(b) Conversely, in certain circumstances one might have very good reasons for thinking it quite probable antecedently that a certain roulette-wheel would *not* be honest. In all artificial cases an essential part of one's ground for holding any reasonable opinion on the antecedent probability of the machine being honest or being biased is knowledge of the general laws of human motivation and of the characters and motives of certain particular individuals. Again, an essential part of one's ground for inferring, from the supposed construction of the machine and its observed performance up to date, to any conclusion about its future behaviour in any assigned respect, is one's knowledge of the general laws of physics and of the properties of specific kinds of matter.

(c) It might therefore seem that there is a risk of *circularity* in taking, as a model for the inductive inference of natural uniformities from observed regularities of co-existence or of sequence, the case of inferring from the past results of spinning a roulette-wheel to the probable results of further spins. I mention this *appearance* of circularity only in order to say that I do *not* think it harmful for the purpose for which the analogy is used. That purpose is simply to exhibit the presuppositions of an inductive argument in a case where they are very obvious, and to suggest (α) that inductive generalization *everywhere* presupposes the finite antecedent probability of something *analogous to bias* in the case of a roulette-wheel or a die, and (β) that this *always* rests on some view about the 'concealed structure and mechanism' (to use those words very widely) of nature as a whole or of a particular department of it.

(5) *Induction by Simple Enumeration and the Hypothetical Method.* Under this heading I will discuss a number of inter-related points raised by Professor von Wright.

I alleged that induction by simple enumeration (so far as it is exemplified by taking counters out of a bag, noting their colours, and then drawing conclusions with more or less probability as to the original proportion of counters in the bag) is a particular case of the hypothetical method. Professor von Wright objects to this. I think that his objection rests partly on a mere difference in the use of words, and partly on an important matter of principle.

(i) The matter on which I think there is no real difference is this. Let $h_0, h_1, ..., h_n$ be a set of mutually exclusive and collectively exhaustive alternative propositions, which it is proposed to test by specific experiment or observation. Let f be any relevant data which one may have *before* undertaking the test, and let Q_N be a summary of the relevant information that has accumulated at the N-th stage of carrying out the test. Then for any typical one of these alternatives h_r the probability relative to the conjunction of f with Q_N is given by the equation

$$h_r|f \& Q_N = [(h_r|f) \times (Q_N|f \& h_r)] \div [\sum_{r=0}^{r=n} (h_r|f) \times (Q_n|f \& h_r)]$$

where any symbol of the form '$p|q$' stands for the probability of the proposition p given the proposition q.

Now in the case of bag-problems the propositions of the form h_r are alternative 'hypotheses' to the effect that exactly so many of the n counters in the bag are of such and such a colour. The proposition Q_N is a summary, at any given stage of the experiment, of the accumulated information as to the whiteness or non-whiteness of the counters drawn and inspected up to that point.

In what is commonly called 'the hypothetical method' we use what is *in principle* the same formula, but there are the following important differences in detail. (a) Instead of considering a number of mutually exclusive and collectively exhaustive alternative propositions $h_0, h_1, ..., h_n$, we consider just a *single* proposition H and its *logical contradictory*

291

$\bar{H}$. (b) H is such that at every stage $Q_N|f\,\&\,H$ is either 0 (in which case the hypothesis is *refuted* and the experiment comes to a natural end), or 1 (in which case there is no reason why the experiment should not be continued). (c) In the bag experiment H is analogous to the single alternative h_n, viz. that *all* the counters in the bag are white. And $Q_N|f\,\&\,H$ is either 0 (if Q_N includes the information that at least one *non-white* counter has been drawn), or 1 (if it consists of the information that *all* the counters drawn up to that stage have been *white*). The formula therefore reduces to

$$H|f\,\&\,Q_N = (H|f) \div [(H|f) + (\bar{H}|f) \times (Q_N|f\,\&\,\bar{H})].$$

So what I was trying to say could be more accurately expressed as follows. The reasoning in induction by simple enumeration (so far as this is accurately represented by experiments in drawing counters from a bag), and the reasoning in the hypothetical method, are instances of essentially the *same general formula* in the calculus of probability. And the latter can fairly be regarded as in certain respects a *more restricted* case of that formula, since it is by definition subject to the three conditions stated above.

(ii) The important difference in principle is this. Is the kind of hypothesis which is tested in what is ordinarily called the 'hypothetical method' *really* on all fours with the $(n+1)$-th of the alternative 'hypotheses' which are tested in an artificial experiment with counters in a bag? Is *All swans are white* a proposition of the same logical kind as *All the n counters in the bag are white?* Professor von Wright objects that the former are propositions about what he calls 'open classes' and that the latter are about 'closed classes', and that these two are fundamentally dissimilar kinds of proposition.

I think that he is right to object, and that I was wrong to overlook this distinction, but that his objection hardly goes far enough. It seems to me now that we have to contrast at least *three* fundamentally different kinds of proposition, (a) 'All S is P' might express simply the proposition that S_1 is $P\,\&\,S_2$ is $P\,\&\,\ldots\,S_n$ is P, and that these are all the S's that there are. (b) It might express a rather complicated proposition of the following form. Consider a sequence of collections of the following kind, viz., $(S_1), (S_1\,\&\,S_2), \ldots, (S_1\,\&\,S_2\,\&\ldots S_n), \ldots$. Let the percentage of the members of these collections which are P be respectively $p_1, p_2 \ldots, p_n, \ldots$. Then 'All S is P' might be taken to mean the same as p_n tends to the limiting value 100% as *n* tends to infinity.' This latter sentence is itself a highly condensed expression for a rather complicated proposition, but we need not unpack it further here. (c) 'All *S* is *P*' might be taken to mean that in the actual world (though not in all possible worlds) *any* instance of *S* *would be* an instance of *P*. I do not know how to analyse such propositions further. But I can perhaps indicate their peculiarity by remarking that one is tempted to say of any such proposition (α) that, if it is true, it is *necessary*, but (β) that the fact that it is necessary is *contingent*. (In contrast with this, one can say of the necessity of a true *a priori* proposition that its belonging to that proposition is itself a *necessary* fact.)

We might call these respectively the 'enumerative', the 'limiting-frequency', and the 'nomic' interpretations of such a sentence as '*All S is P*'. It is immediately obvious that (b) differs from (a). If it is not immediately obvious (as I think it should be) that (c) differs from (b), this becomes evident when one reflects that (b) is compatible with there being any finite number of *S*'s which are *not P*, whilst (c) is not compatible with there being a single *S* which is not *P*.

Now the 'limiting-frequency' interpretation certainly presupposes 'open classes', in the sense of classes which contain an infinite number of members. For that is involved in the notion of a limit. For that very reason I doubt whether it has any application outside pure mathematics. The 'nomic interpretation' does not presuppose 'open classes' in that sense. For the proposition that *any* instance of *S would* be an instance of *P* in the actual world is consistent with the number of actual instances in the whole course of the world's history being finite or even zero. What it does presuppose is the notion of classes determined by *intension* as distinct from by *enumeration* of their members.

It seems to me that what we commonly try to test by the so-called 'hypothetical method' is universal propositions in the *nomic* sense. If so, they *are* fundamentally different from such propositions as *All the n counters in the bag are white*. But the difference is even more fundamental than would be suggested by the contrast between 'open' and 'closed' classes.

(6) *The Theory of 'Generators'*. I have very little to object to in Professor von Wright's comments in what I said about this in *P. of P.I.*

(i) He is correct in saying that the argument on p. 27 of that paper[5] does not presuppose that the number of *generated* characteristics is finite. It presupposes only that n, the number of *generating* characteristics, is finite. The further argument, in the section entitled *Effect of the Relative Values of n and N* certainly assumes N to be finite when considering the alternatives that N is less than or equal to n, since n is assumed throughout to be finite. In discussing the alternative that N is greater than n, I certainly did assume in my own mind that N is finite; and, although the mere supposition that N is greater than n does not entail this, there are many steps in the argument which presuppose it.

(ii) He is correct also in saying that I have nowhere shown that the factor $(\mu_r \& v_s)|h$ in the formula on p. 27 is greater than 0.6 This is the probability (relative to the general assumption of the theory of generators, and to the special assumption that each generated characteristic is generated by *only one* set of generators) of the proposition expressed by the sentence: 'In a generalization, whose subject is a conjunction of μ generated characters, and whose predicate is a conjunction of v generated characters, the former require exactly r, and the latter exactly s, generating factors respectively to generate them.'

(iii) He says, rightly, that all my arguments presuppose that the antecedent probability of a generalization 'can be linked with a ratio of true generalizations among a class of generalizations'. But he complains that the nature and justification of this link are not made clear. I do not see exactly what the difficulty is here. If it could be shown that at least a certain proportion of possible generalizations of a certain kind *must* be true, *e.g.*, at least $p\%$ of generalizations with a μ-fold subject and a v-fold predicate, surely the antecedent probability of any generalization of that kind would be at least $\dfrac{p}{100}$.

(iv) He mentions my remarks on p. 41 of *P. of P.I.*,[7] that the generating factors must be supposed to be *determinable* characters, and that it would follow that the generated characters must be so too. He finds the notion of 'determinables' and 'determinates' obscure, and asks me to try to clarify it.

I regret that it is impossible for me to go into this very large question here. The following very sketchy and therefore rather obscure remarks must suffice. (a) My account of generating factors explicitly assumes that no conjunction of such factors is either logically necessary or logically impossible. The statement about generating factors having to be determinable characters is bound up with this. (b) That is because of the following properties of determinable characters and of determinate characters. Supreme determinables are all logically independent of each other. But it is logically necessary that any thing which possesses a determinate character should possess all the determinables, of whatever order, under which this falls. And it is logically impossible that any thing should possess two determinate characters of the same order which fall under one and the same determinable.

I *think* that a more accurate statement of what I had in mind would run as follows. A complete collection of generating factors would have either (a) to contain nothing but *supreme determinables*; or (b) to contain nothing but *determinates*, each of which falls under a *different* supreme determinable; or (c) to be a *mixture* of (α) supreme determinables, and (β) determinates, none of which fall under any of these determinables, and each of which falls under a different supreme determinable.

(7) *Necessary Conditions and Sufficient Conditions*. I take some credit for seeing by 1930, when I published my two papers on 'Demonstrative Induction,[8] that these are the essential concepts involved in demonstrative induction, and for having worked out the formal logic of them in some detail and without serious mistakes, though not without one very serious omission. But all that I have written on this topic has now been superseded by Professor von Wright's more thorough and more accurate work.

I agree with him that it sounds odd to say: 'The ground becoming wet is a necessary condition of rain having fallen in the neighbourhood', and I agree that both he and I are committed by our definitions to saying such things. I agree too that the verbal paradox is bound up with the conviction that a *causal* condition must be fulfilled *before* that which it conditions begins. That is why, in the present essay, I have introduced the terms 'necessary *precursor*' and 'sufficient *precursor*', when discussing, in Section IV, C, 2 above, Professor Russell's comments on my account of Causation in the *Examination of McTaggart's Philosophy*.

I should be inclined to say that we must distinguish between a 'condition', in the sense of a *ground for inference*, and a 'condition' in the sense of a *factor in causation*. Given a knowledge of causal laws, one can often infer from knowledge of a *later* event to the conclusion that such and such an *earlier* event must have happened. (Unless one is a prophet, one cannot of course infer from knowledge of a *future* event to the occurrence of such and such an event in the *present or the past*, since one cannot be in possession of such knowledge.) But, when a person makes such an inference from a later to an earlier event, he does so because he has reason to believe that the later state of affairs (*e.g.*, the ground being wet) would have come into being *only* if it had been preceded by a state of affairs containing such and such an event (*e.g.*, a fall of rain in the neighbourhood) as a cause-factor. [...]

NOTES

[1] *Editor's note:* This chapter is reprinted from P. A. Schilpp, editor, *The Philosophy of C. D. Broad* (The Library of Living Philosophers, 1959), pp. 741–764. It constitutes a part of Broad's comments on the essays included in Prof. Schilpp's volume. In the part which is reproduced here, Professor Broad discusses, in addition to Professor von Wright's essay (reprinted in the present volume), also an essay by L. J. Russell entitled 'Substance and Cause in Broad's Philosophy' (*op. cit.* pp. 263–280) and one by Everett J. Nelson entitled 'Some Ontological Presuppositions in Broad's Philosophy' (*op. cit.* pp. 71–93). Although these papers do not deal with induction and probability directly, many of Broad's comments on them are of interest here. It is hoped that Broad's answer to Professors Russell and Nelson are intelligible although their papers are not reproduced here.

[2] *Editor's note:* Professor Broad's answer to Hanson is omitted here.

[3] *Editor's note:* Present volume, pp. 159–183.

[4] *Editor's note:* Present volume, pp. 1–52 and 86–126, respectively.

[5] *Editor's note:* Present volume, pp. 107–108.

[6] *Editor's note:* Present volume, p. 108.

[7] *Editor's note:* Present volume, p. 118.

[8] *Editor's note:* Present volume, pp. 127–158.